The Four Corners of Mathematics

The Four Corners of Mathematics: A Brief History, from Pythagoras to Perel-man describes the historical development of the 'big ideas' in mathematics in an accessible and intuitive manner. In delivering this bird's-eye view of the history of mathematics, the author uses engaging diagrams and images to communicate complex concepts while also exploring the details of the main results and methods of high-level mathematics. As such, this book involves some equations and terminology, but the only assumption on the readers' knowledge is A-level or high school mathematics.

Features
- Divided into four parts, covering Geometry, Algebra, Calculus and Topology
- Presents high-level mathematics in a visual and accessible way with numerous examples and over 250 illustrations
- Includes several novel and intuitive proofs of big theorems, so even the non-expert reader can appreciate them
- Sketches of the lives of important contributors, with an emphasis on often overlooked female mathematicians and those who had to struggle.

Thomas Waters was born and grew up in Dublin, Ireland, completing his undergraduate degree and PhD in Mathematics at Dublin City University. This was followed by a post-doc at the University of Strathclyde in Glasgow, then three years as a lecturer at the National University of Ireland, Galway, before becoming a lecturer at the University of Portsmouth, England, in 2010. During his time Thomas has lectured on very many different topics such as Linear Algebra, Differential Geometry, Knot Theory and Analytical Mechanics, to classes large and small, from one student to 300. His research interests go from General Relativity (stability of naked singularities) to Astrodynamics (solar sails in the 3-body problem) to Riemannian Geometry (integrability and the conjugate locus), but a unifying theme would be ordinary differential equations and periodic solutions. Thomas has two daughters, and when he is not being their taxi driver, he and his wife enjoy long walks in the woods and mountains.

AK Peters/CRC Recreational Mathematics Series

Series Editors

Robert Fathauer
Snezana Lawrence
Jun Mitani
Colm Mulcahy
Peter Winkler
Carolyn Yackel

Mathematical Conundrums
Barry R. Clarke

Lateral Solutions to Mathematical Problems
Des MacHale

Basic Gambling Mathematics
The Numbers Behind the Neon, Second Edition
Mark Bollman

Design Techniques for Origami Tessellations
Yohei Yamamoto, Jun Mitani

Mathematicians Playing Games
Jon-Lark Kim

Electronic String Art
Rhythmic Mathematics
Steve Erfle

Playing with Infinity
Turtles, Patterns, and Pictures
Hans Zantema

Parabolic Problems
60 Years of Mathematical Puzzles in Parabola
David Angell and Thomas Britz

Mathematical Puzzles
Revised Edition
Peter Winkler

Mathematics of Tabletop Games
Aaron Montgomery

Puzzle and Proof
A Decade of Problems from the Utah Math Olympiad
Samuel Dittmer, Hiram Golze, Grant Molnar, and Caleb Stanford

A Stitch in Line
Mathematics and One-Stitch Sashiko
Katherine Seaton

The Four Corners of Mathematics
A Brief History, from Pythagoras to Perelman
Thomas Waters

For more information about this series please visit: https://www.routledge.com/AK-Peter-sCRC-Recreational-Mathematics-Series/book- series/RECMATH?pd=published,forthcoming &pg=2&pp=12&so=pub&view=list

The Four Corners of Mathematics

A Brief History, from Pythagoras to Perelman

Thomas Waters

University of Portsmouth, UK

CRC Press
Taylor & Francis Group
Boca Raton London New York

CRC Press is an imprint of the
Taylor & Francis Group, an **informa** business

AN A K PETERS BOOK

First edition published 2025
by CRC Press
2385 NW Executive Center Drive, Suite 320, Boca Raton FL 33431

and by CRC Press
4 Park Square, Milton Park, Abingdon, Oxon, OX14 4RN

CRC Press is an imprint of Taylor & Francis Group, LLC

Library of Congress Cataloging-in-Publication Data
Names: Waters, Thomas (Mathematician) author.
Title: The four corners of mathematics : a brief history, from Pythagoras to Perelman / Thomas Waters, University of Portsmouth, UK.
Description: First edition. | Boca Raton : AK Peters/CRC Press, 2025. | Series: AK Peters/CRC recreational mathematics series | Includes bibliographical references and index.
Identifiers: LCCN 2024029169 (print) | LCCN 2024029170 (ebook) | ISBN 9781032596518 (hbk) | ISBN 9781032594989 (pbk) | ISBN 9781003455592 (ebk)
Subjects: LCSH: Mathematics--History--Popular works. | Geometry--History--Popular works. | Algebra--History--Popular works. | Calculus--History--Popular works. | Topology--History--Popular works.
Classification: LCC QA21 .W328 2025 (print) | LCC QA21 (ebook) | DDC 510.9--dc23/eng20240806
LC record available at https://lccn.loc.gov/2024029169
LC ebook record available at https://lccn.loc.gov/2024029170

ISBN: 978-1-032-59651-8 (hbk)
ISBN: 978-1-032-59498-9 (pbk)
ISBN: 978-1-003-45559-2 (ebk)

DOI: 10.1201/9781003455592

Typeset in Latin Modern
by KnowledgeWorks Global Ltd.

Publisher's note: This book has been prepared from camera-ready copy provided by the authors.

To Maria, Emily and Rachel
The three best things that ever happened to me

Contents

PART IV Topology

Preface

When I was in the final year of my undergrad maths degree, I (like many students) had grown jaded from years of lectures and exams and I was starting to forget why I chose to study maths in the first place. One day I was in the library looking for a book, I have long forgotten which one, but when I found it on the shelf it was the book just beside it that caught my eye. It was *Euclidean and non-Euclidean geometries; development and history* by Marvin Greenberg (see number [51] in the bibliography), and the cover had this bizarre image of angels and demons receding into some sort of symmetrical infinity. I took that book home instead of the other, and it blew my mind. I had forgotten how rich and complex how elegant and deep mathematics was, but also how maths is a collaborative effort gradually built up over millennia, one step forward leading to the next. I hope in this book to give you, the reader, some sense of how creative and inspiring mathematics can be, by sketching for you some of the big ideas in mathematics over the centuries and the people who had them, but first, a disclaimer.

My wife and I enjoy a TV show where a group of candidates are gradually whittled down to a lucky handful who get to present their business plans to some tetchy tyrant. We are always amazed at who manages to avoid the chop, and we reckon it is because the producers already know the business plans and have decided who is going to make it through to the end. In fact this comes up so often we just refer to it as 'Statement A', and every time some hopeless chancer makes it through to the following week we look at each other knowingly and say 'Statement A'. I am going to do something similar here:

Statement A: This book does not attempt to be comprehensive.

Mathematics is a vast and complicated beast, perhaps the most elaborate creation of the human mind, I couldn't hope to cover it all; and besides, there are already lots of really excellent comprehensive 'history of maths' books out there. I would recommend Katz [68], Stillwell [109] or Boyer [16], and I will cite several others as we go along. This book does not attempt to replace those ones, instead my aim is to give a whistle-stop tour through the development of mathematics by focusing on the 'big ideas' that lead to major new branches of maths developing. My hope is that the treatment will be accessible, by which I mean the content could be readily understood by someone who is simply interested in science in general, and to that end I will try to avoid too much technical detail and jargon and will instead focus on description, suggestion and visualization. Having said that, if you already have a high level of mathematical knowledge I hope you will find some of the proofs and descriptions presented herein novel.

Mathematics is the tree with many branches. In this book I will focus on just four: Geometry, Algebra, Calculus and Topology, and I will try to describe how each grew from early ideas and then split into separate specialized fields as it matured. This is as opposed to a chronological approach [114] where, for example, I would describe everything that happened in the 18th century and then everything that happened in the 19th century and so on. Having said that, I do try to have some element of chronological development within the four branches, and the order of the branches themselves is arguably chronological (see Figure C.1). As is natural I have chosen to focus on topics that I find interesting and which suit the level of this book; that are important for the narrative development and will be needed later in the text; and which should perhaps be on an undergraduate course but are not because of an emphasis on assessments and graduate prospects. I am sure every mathematician will flick through the table of contents and say "but you didn't include...!" and to them I say: Statement A.

Though traditionally mathematics was seen as male, pale and stale (*Men of Mathematics* [11] would be a good example), recently there is much commendable effort to diversify and decolonize mathematics (such as [69] and [14]). I will attempt to contribute to that in my own way, for example there are only images of female mathematicians in this book; I think the world has seen enough pictures of Newton and his flowing locks (you know the one I mean). While this may not be *Ulysses*, the style of the chapters can vary from a more direct treatment of a specific result or theorem, to a broader discussion of the lives and interactions of the main players in the development of a topic, to some combination of the two (usually the first chapter in each part is more of a broader historical sketch). I am not a comedian but I will occasionally try and lighten things with a joke here or there; if I cross the line into facetiousness please forgive me. I must also apologise in advance if I sometimes treat the formation of ideas in an overly simplistic way; the fact is that every time you think a specific result or idea is due to a certain person there is always someone who had a rougher earlier version of it: Euler's polyhedron formula? Descartes had it 100 years before. Pascal's triangle? The Chinese had it 400 years before. Pythagoras theorem? The Babylonians knew it 1000 years before! So while my treatment might seem limited in places I will attempt to provide references for further reading where I can. I will also include some references to websites because there's some great stuff out there: imagine you could watch a YouTube video of Riemann's *habilitationsschrift* lecture; imagine Newton publishing on TikTok 'hey guys, your buddy Zak here, I was sitting under a tree just now. . . ' The spellings and accent marks on some names can change from one source to the next, so I will try to follow the most recent conventions (and will randomly add or not the possessive 's' to the end of names that end in s). All images (except where stated otherwise) were drawn by myself mostly using Mathematica; all text (except direct quotes) was written by myself and not a robot (although that is what a robot would say).

Finally, I wanted to include several equations, not because we will step through logical chains of reasoning to prove something (although we will do that in places) but because mathematical symbols are in themselves beautiful and elegant, like some secret magical language that only people like you and I, the initiated, can understand.

I

Geometry

CHAPTER 1

The Beginnings

T HE DAWN doesn't happen in an instant. It doesn't go from night to day from one moment to the next; instead there is a long hazy twilight that gradually takes shape as morning. In the same way Mathematics didn't begin with the Greeks; Egyptian and Babylonian scribes knew how to solve linear, quadratic and cubic equations as we will see, and the Babylonian *Plimpton 322* tablet from 1800 BCE has tables of Pythagorean triples (see Figure 5.1). Indeed humans were carving tally-marks into animal bones at least 20,000 years ago (see for example [68] or [16] for a detailed description of pre-Greek mathematics). However, 'Mathematics' in the modern sense: formal, generalized and abstract, with definitions and logical deduction, had its first (documented) hurrah in Classical Greece and so that is where we will start this chapter.

And where better to start than with Pythagoras? His theorem needs no introduction, being probably everyone's first experience of 'real' maths as a young person. Pythagoras established a school in the Greek colony of Croton in southern Italy about 530 BCE. Mysterious and magical, he was more like a mystical cult leader and philosopher than a rational thinker. He preached the existence of the soul and reincarnation; he was a vegetarian (and yes forbade his followers from eating beans [60]); the Pythagoreans had female members, initiation rituals and vows of secrecy. There were two aspects to the Pythagorean movement: *akousmatikoi* (rules describing how to live well: silence and abstinence, moral behaviour to guarantee a good afterlife) and *mathematikoi* (the mystical attempt to understand the fundamental structures of the universe through number and harmony). 'Mathema' here means knowledge in general, for example in words like polymath, so in this sense at least the Pythagoreans were certainly the first 'mathematicians'. They saw

DOI: 10.1201/9781003455592-1

Figure 1.1: *Pythagoreans celebrate the sunrise* (1869) by Fyodor Bronnikov

geometry and number as a way of connecting with the fundamental nature of the world, of discovering universal truths to be closer to God. Already the Pythagoreans were aware of the work of Thales who (it is claimed) had proved that the angle in a semi-circle is a right angle, the two base angles in an isosceles triangle are equal, and the three angles of a triangle add to two right angles. It is worth seeing this last statement, which I will refer to as 'Thales theorem' [35], through the Pythagoreans' eyes: we are saying not just that this triangle or that triangle has a certain property, but *all* triangles, no matter how big or small or where you draw them or imagine them to be; this is a *universal* property of triangles. Also if the right angle is what other angles are compared to, how marvellous that the three angles of a triangle should add to precisely *two* right angles, not a bit more or a bit less but precisely two. There is an wondrous simplicity here, like a gift from the Gods; imagine how hard life would be if the three angles of a triangle added to 2.0012312...right angles. These two notions, universality and simplicity, inspired the Pythagoreans and have been inspiring mathematicians ever since. But it is more than that: Thales proved that the three angles of a triangle add to two right angles and this fact is as true now thousands of years later as it was then, and it will still be true in thousands of years' time, but it was *always* true, even before Thales proved it, we just didn't know! He had uncovered a hidden fundamental truth, and this is the timeless aspect of knowledge that the Pythagoreans sought.

But enough of that, let's get to Pythagoras theorem. Nowadays we see it in the following way: if you have a right angled triangle with long side of length c and the other two sides of length a and b, then

$$c^2 = a^2 + b^2.$$

We would say "the square of the hypotenuse equals the sum of the squares of the other two sides", mostly because we use it today to calculate lengths. But this is not how the Pythagoreans would have seen it; firstly they did not have symbols and equations (that came a good 2000 years later, as we will see in Part II), but also they did not think in terms of lengths of sides but rather the areas of squares drawn on the sides; they would say "the square *on* the hypotenuse equals the sum of

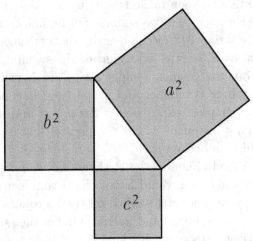

the squares *on* the other two sides", as in the *areas* of those squares (see Figure above). It is not known how Pythagoras proved this theorem or indeed whether he even proved it at all, but it has been proven many times ([77] has 370 proofs!) and here is one Pythagoras would probably have appreciated:

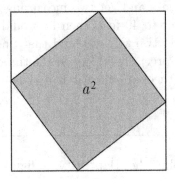
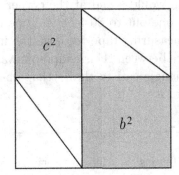

The two large squares on the left and right have the same area, and the empty triangles you can see (4 on the left, 4 on the right) are copies of the right angled triangle we are interested in. The gray square on the left is the square on the hypotenuse, and the two gray squares on the right are the squares on the two short sides of the triangle. Since the gray areas on the left must equal the gray areas on the right (each being the big square minus the 4 triangles), the square on the hypotenuse is equal to the sum of the squares on the other two sides.

The Pythagoreans were hugely influential. For example, they observed that if you pluck a stretched string of a certain length to make a note, and then make the string precisely half that length and make another note, the two sounds will be harmonious to our ears. If you make the string 2/3 of the length, another harmonious sound comes. Simple ratios of lengths led to harmonious sounds, and this was like a message from the cosmos: simple ratios are harmonious and pleasing, and everything is number and proportion. It is hard to resist this feeling that the simplest expression is the truest and most natural, and we will see many instances where complex setups resolve into surprisingly simple rules and relations.

And of course the Pythagoras theorem itself is tremendously useful, although you would have a hard time convincing people of that. When I was a student I worked on the marquees, those large tents people put up for weddings and parties. One of the guys I worked with (nice bloke, prison tattoos on his knuckles, I didn't argue with him much) heard I was studying maths in university and said to me "sure when will I ever need to use the bleedin' Pythagoras theorem?" And I was able to reply "you literally used it this morning!" The first thing we do in laying out the tent is position the plates that the legs of the tent will sit on, and we want the sides of the tent to make a nice right angle in the corner. So the foreman would stand at the corner plate and get two measuring tapes, stretch one out to 30 feet and one out to 40 feet; then he would use a third measuring tape to move the first two this way and that until the distance between their endpoints was precisely 50 feet. If the three sides of a triangle satisfy the Pythagoras theorem, then it is a right angled triangle.

If Thales is the mysterious John the Baptist-type figure, and Pythagoras the visionary who started a mystical movement that lost its way after he died, then Euclid is the St. Paul who codified and put a logical structure on everything that had gone before him. A scholar in the Library of Alexandria in northern Egypt, Euclid wrote his master-piece *the Elements* [30] in about 300 BCE, and it was hugely significant for two reasons:

Firstly *the Elements* brought together all the mathematics that was known at the time into one place, spanning 13 books with numerous definitions (131 altogether) and many theorems called 'propositions' (465 in total). Each book had a theme, for example Book I is about lines and triangles (Pythagoras theorem is I-47, see Figure 1.4; its converse, which is what we really used in laying out the tent, is I-48), Book III is about circles ('the angle in a semicircle is a right angle' is III-31) and Book XIII is about regular solids such as the tetrahedron and dodecahedron (which in a pleasing coincidence we will see in Chapter 13 of this book).

The other, and perhaps more important, aspect of *the Elements* is the so-called 'axiomatic' approach Euclid adopted. Consider this situation: suppose we want to prove Thales theorem that the three angles in a triangle add to two right angles. The standard approach (I-32) is as follows: consider the triangle ABC. We draw a line segment PQ par-allel to one side, let's say the side BC, through the opposite vertex, A. Since the angle PAB equals ABC, the angle QAC equals ACB and the angles PAB, BAC and QAC add to two right an-gles, this means ABC, BAC and ACB must add to two right an-gles; done. But the Platonic school Euclid came from would question

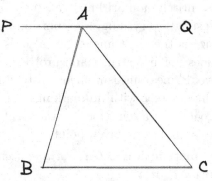

this: how do we know PAB equals ABC? How do we know we can draw a line parallel to any side through an opposite vertex, and is there only one parallel line to choose from? What even is a parallel line or worse: what is a line? At some point, we must accept certain terms and statements as fundamental and use them then to define and prove other terms and statements; the trick is to begin with what seems absolutely irrefutable, that everyone can agree on. Once you have that, everything you build on top of it will be rock-solid.

To this end Euclid started Book I with 23 definitions (what is a 'point', what is a 'line', what do 'perpendicular' and 'parallel' mean, etc.), 5 common notions (for example *things which equal the same thing also equal one another*) and 5 axioms, also known as postulates. These postulates are the foundations of the Geometry that Euclid then went on to set forth, which is known as Euclidean Geometry in his honour. They are (I am paraphrasing slightly):

1. A line segment can be drawn from any point to any other point,

2. A line segment can be produced continuously in a straight line,

3. There exists a circle with any given center and radius,

4. All right angles are equal to one another,

5. Given a line l and a point P not on l, there exists one line through P parallel to l.

This last postulate, known as 'the parallel postulate' or 'Euclid's fifth', is the nightmare rabbit-hole that generations of mathematicians fell into; we will come back to it in Chapter 2.

Showing the influence of the Pythagorean school, Euclid's *Elements* is entirely non-arithmetical: there are no numbers and no measurement. Instead everything is proportion and ratio, and the only angle of fixed size is the right angle. Euclid uses 'constructions' in his proofs, but he does not use a ruler and protractor to draw geometrical figures as this would mean measurement; instead he uses a compass and 'straightedge' (like a ruler with no markings), see [57]. Let us work toward a modest goal to get a sense for how the great mind thought: to draw a line tangent to a circle at a given point.

Construction 1 *Given a line segment AB, to construct an equilateral triangle on it.*

This is actually the very first proposition in *the Elements*, I-1, and it is simplicity itself: put the 'spike' of the compass on A, the 'pencil' on B, and sweep out a circle (which we know exists from Postulate 3). Do the same again by putting the spike on B and the pencil on A. The two circles will intersect at two points and, choosing one, join it to A and B with line segments (which we know we can do from Postulate 1). You have an equilateral triangle. □

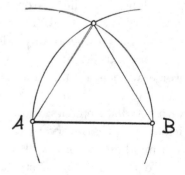

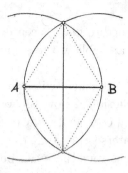

Construction 2 *Given a line segment AB, draw a line which bisects it perpendicularly.*

We construct two equilateral triangles on the segment AB which we know we can do from Construction 1; then we join their two vertices which we know we can do from Postulate 1; this segment bisects AB, and the angles formed are right angles. □

Construction 3 *Given a circle whose center O is known, and a point P on that circle, draw the line tangent to the circle at P.*

We put the spike of the compass on P and the pencil on the center O and sweep out a circle (Postulate 3). Then we extend the segment OP until it reaches this circle at the point C (Postulate 2). Then we draw the perpendicular bisector of OC (Construction 2); this line is then tangent to the circle at P. □

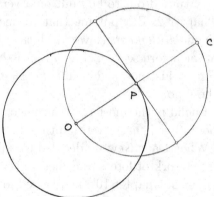

See how this works? Rather than jumping straight into drawing the tangent line, we instead break the process into its *elements*, which we can then use for more complicated constructions. Notice also how careful we need to be in the language used, for example in the last construction we explicitly said we knew where the center of the circle was; if not, we can use the following construction:

Construction 4 *Given three points A, B, C that don't lie on a line, construct the circle that passes through these three points.*

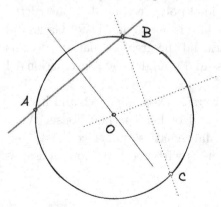

We join any two pairs of the three points with line segments (Postulate 1), construct the perpendicular bisector to those line segments (Construction 2), and continue them until they intersect at O. This is the center of the circle which passes through the three given points. □

To draw the circle in the last construction we first had to find its center, so if we were presented with a circle and needed to find its center we could simply choose three points on the circle and follow the same steps.

What we have just walked through is a veritable minefield with several hidden assumptions. For example, in Construction 4 how do we know the perpendicular bisectors will intersect when extended? In Construction 1 how did we know the two circles will intersect, or that they will intersect in just two points? To be absolutely rigorous there can be no ambiguity, no hand waving, and flaws in Euclid's presentation were pointed out even in Classical times; in fact the whole of Euclid's geometry turned out to be built on a very subtle quicksand, as we will see in Chapter 2. But putting that to one side for now, we must accept that Euclid's *Elements* elevated Mathematics to a new level of sophistication never before seen, a singular and shining accomplishment. It is hard to convey just how significant *the Elements* was and is, but let me try with these points:

- Euclid is the most read author of all time. The Bible is the most read book, but it has several authors; *the Elements* has only one: Euclid.

- When Isaac Newton decided to unleash on the world his mathematical framework of forces and mechanics in the Principia (which we will talk about in Chapter 10), he reached for the gold standard: he presented his work in the style of *the Elements*. He began with definitions, common notions, postulates and then proceeded in many propositions. I would argue that the *Principia* was the finest expression of Classical Greek mathematics, the zenith of the Euclidean program. In fact I would go so far as to say that, apart from getting used to the notion of 'force', Euclid would be very comfortable reading the *Principia*.

- When Gauss, one of the giants of mathematics who we will hear mention of many times in this book, was cutting his teeth, his first major contribution was to show that a 17-sided polygon could be constructed with straightedge and compass. The point is not so much that Gauss did this, but that even 2000 years after Euclid the greatest mathematicians of the day still respected the compass and straightedge constructions of *the Elements*.

- My wife's great-grandfather sat university exams in 1882, and his certificate is proudly displayed on the wall of her parents' house. What subject does it say he studied? Not 'Mathematics', not 'Geometry', but 'Euclid'. Euclid was literally synonymous with geometry for thousands of years.

After Euclid, Mathematics continued to grow more and more elaborate in the hands of the Greek masters, and this book will not attempt to describe all their contributions, however there are two names that stand out: Archimedes (who we will say more about in Chapter 9) and Apollonius of Perga who we shall discuss presently. As the sophistication and influence of Mathematics spread through the Greek world, we see both original works (such as Euclid's *the Elements* and Archimedes' *Quadrature of the parabola*), but also 'commentaries', which are rather like textbooks we would see today: a presentation of well known results by another author, with discussion points and elaborations. Indeed many classical texts only survived to the modern age through commentaries, a good example being the commentary on Apollonius' *Conics* by Theon of Alexandria, which was written also by his daughter, Hypatia.

Hypatia of Alexandria (c. 360-415 CE) was a renowned philosopher, astronomer and mathematician, a woman *who made such attainments in literature and science, as to far surpass all the philosophers of her own time* [15]. Alexandria was at the center of the Hellenic intellectual world (or at least on a par with Athens), and students from all over the Mediterranean came to Alexandria to be taught by scholars such as Hypatia. We have no images of her of course but Figure 1.2 is from a similar time and place [61]. It is said that she wore the cloak of the philosopher, and is described as having been very beautiful (although isn't it odd that female mathematicians tend to have their appearance commented on? Nobody wrote "Euclid was a brilliant geometer and by all accounts he was a bit phwoar!"). She wrote commentaries on Ptolemy's *Almagest* (150 CE), Diophantus' *Arithmetica* (250 CE), and Apollonius' *Conics* (225 BCE).

Figure 1.2: A *Fayum* portrait from northern Egypt from a period and culture similar to that of Hypatia.

The conic sections were in fact known before Apollonius, but, like Euclid and *the Elements*, his treatment of the subject in *Conics* superseded all others and stands beside *the Elements* and the works of Archimedes as the purest expressions of Greek mathematics [16]. Apollonius described the conic surface more generally than his predecessors, in fact Definition 1 of *Conics* defines them so [9]: take a circle and a point not in the plane of the circle (the 'vertex'). Draw a line from the vertex to any point on the circle, and sweep that line around all the points on the circle while still passing through the vertex; the surface swept out is a conic surface or cone.

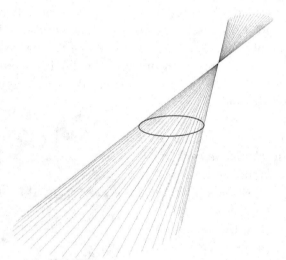

Nowadays we would describe this as an 'oblique' cone. Apollonius then went on to consider planes that intersect the cone, and coined the terms for the three curves which result: the parabola, the hyperbola and the ellipse. The ellipse is the curve we will see repeatedly in this book, and looking at Apollonius' definition of an ellipse (Book I Proposition 13) we can see how important was the role of interpreters like Hypatia: *If a cone is cut by a plane through its axis, and is also cut by another plane which on the one hand meets both [lateral] sides of the axial triangle, and on the other hand, when continued, is neither parallel to the base [of the cone] nor antiparallel to it, and if the plane of the base of the cone and the cutting plane meet in a straight line perpendicular either to...* (the English translation of this sentence has over 250 words, I've never made it all the way through)*. . . the diameter is cut off by the straight line drawn from the section to the diameter, this plane is [the rectangular plane under two mentioned straight lines] and decreased by a figure similar*

and similarly situated to the plane under the mentioned straight line and the diameter. I will call such a section an ellipse. Phew! You will forgive me if I don't quote directly from the source again; a diagram is better:

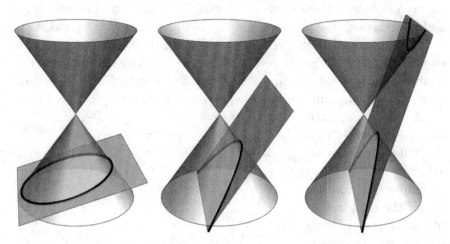

We see there are three different types of curve depending on the angle the plane of intersection makes: when the plane is shallow we have an ellipse (on the left), when the plane is parallel to the side of the cone we have a parabola (in the center), and when the plane is steep it cuts both halves of the cone and so the resulting curve has two components and is known as a hyperbola (on the right). Apollonius proceeded to develop an exceedingly detailed study of these curves, proving numerous complex theorems over 8 books (of which 7 survived) which are rightly seen as one of the most sophisticated achievements of Greek mathematics; we will give here some typical results to get a flavour of Apollonius's work.

The ellipse has two lines of symmetry, the major and minor axes (shown dashed), and the distances from the center to the furthest and nearest points are called the 'semi-major' and 'semi-minor' axes respectively; these describe the shape and size of the ellipse, and when they are equal the ellipse is a circle.

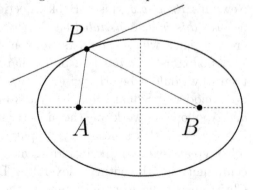

There are two points on the major axis, labeled A and B in the figure, with the property that the sum of the distances from any point P

on the ellipse to A and to B is a constant (III-52). I have also drawn a tangent line to the ellipse at the point P, and III-46 proves that the angle that the segments AP and BP make with the tangent are equal; this means a ray of light shone out of the point A would reflect off the ellipse and pass through the point B, and this is true for a ray of light sent out in any direction. For this reason we call each of A and B a 'focus' or together as 'foci' (there are several 'whispering galleries' which are elliptically shaped rooms; if you stand at one focus and whisper something then a person standing at the other focus will hear it).

A more sophisticated result, VII-31, is the following: take an arbitrary point P on an ellipse (for example one of the dots in Figure 1.3), and draw a line through the center till it meets the ellipse again; this line is called a 'diameter'. Now draw another line through the center, parallel to the tangent at P; this

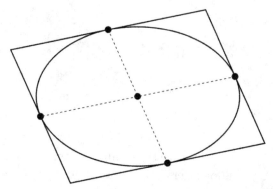

Figure 1.3: Conjugate diameters and their parallelograms (see also Figure 10.2).

is also a diameter, and these two diameters are called 'conjugate'. Now draw tangent lines at the four points where the conjugate diameters meet the ellipse, forming a parallelogram. The theorem is: the area of the parallelogram is the same, wherever you chose the initial point P. In Newton's *Principia*, this is Book I Section II Lemma XII, and as proof he says *this is demonstrated by the writers on the conic sections*; these are the giants whose shoulders Newton stood on.

Apollonius's *Conics* is a pretty hard read to the modern eye, but then why would it be otherwise? We would not expect to be able to pick up a modern graduate text in maths and understand it immediately, and Apollonius's work was the highest level in mathematics at the time. Despite her father's hesitancy, Hypatia taught the *Conics* along with other great texts to all, including members of the growing Christian community in 5th century Alexandria. Despite the official tolerance for Christians in the Roman empire at the time, there was much tension between the bishops and the prefects, and the respect shown to Hypatia by the political establishment made her a target. Rumours were spread that she was in fact a 'sorceress', and in 415 CE *on a fateful day in the holy season of Lent, Hypatia was torn from her chariot, stripped naked, dragged to the church, and inhumanly butchered by the hands of . . . savage and merciless fanatics* [68].

Hypatia has become something of an icon over the years: to some she is a strong independent woman taken down by jealous misogyny; to some she is a shining intellect snuffed out by the ignorant and superstitious; and to others a final hurrah for the Hellenistic worldview before Europe was engulfed by Christian narrow-mindedness. However you view her, while there were some who tried to carry the flame it is clear that her death came at the end of the first golden age of Mathematics. Why did it come to an end? By the time of Hypatia it was already the best part of a thousand years since Pythagoras set up his school in Croton, and nothing lasts forever. Indeed some would say [16] that the real 'golden age' was the four hundred years or so that included Euclid, Archimedes, Apollonius and Ptolemy; the period leading up to Hypatia was more of a 'silver age'. Also some blame must be laid at the feet of early Christianity: in the Hellenistic pantheon there was room for many Gods and immortals, and while I might think my Gods are better than your Gods I would not necessarily deny your Gods exist; it was even tolerated for some to suggest there are no gods at all [131]. In contrast, Christianity was ruthlessly dogmatic and allowed for no questioning or debate; not a good environment for intellectuals. Finally, the real reason might be more prosaic: the mathematics of Euclid and Apollonius was complicated and only understood by a small number of scholars who then passed that knowledge on to the next generation. If that thread was disrupted due to war, oppression or lack of patronage, then soon there would be no-one left who could understand mathematics let alone teach it. If knowledge, for the sake of knowledge, is not nurtured it will die; a lesson from history our political and university leaders should bear in mind.

But Mathematics did not die, it merely moved to the next center of learning, which in the centuries after Hypatia was primarily in the Islamic world. By the 8th century a vast Islamic empire had been established, which then quickly absorbed and developed the culture and science of the peoples they had conquered, just like the Greeks had done when they swept down from the upper Balkans toward the Mediterranean. Though reputed to have burned the books of

Alexandria, early Islamic leaders also set up a 'House of Wisdom' in Baghdad, and called to it the most prominent scholars in the region who translated classical Greek texts into Arabic, such as Euclid's *Elements* (see Figure 1.4) and Ptolemy's *Almagest* (whose name actually comes from 'the greater' in Arabic), but also mathematical texts from India such as Brahmagupta's *Brahmasphuta Siddhanta* [16]. The most significant development from those centuries before the Renaissance was in the new branch of Algebra, and we will tell that story in more detail in the second part of this book. From a geometrical perspective, we see in this period the beginnings of trigonometry.

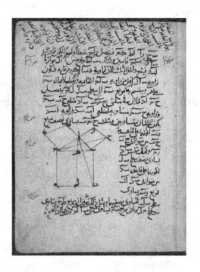

Figure 1.4: Euclid's *Elements* translated into Arabic by Nasir al-Din al-Tusi, 1258 (courtesy of the British Library Board, Add. 23387, f.28).

Already in 1st and 2nd century CE Greece we see people like Menelaus and Ptolemy making the first foray into the measurement (-metry) of triangles (trigon), trigonometry. Their main idea was the chord: draw a line from one point on a circle to another, such as A and B in Figure 1.5, and the length of that chord relates to the length of the arc of the circle between the two points. In the Hindu world however, a more modern viewpoint developed during the 4th and 5th century CE, described in the *Siddhantas* (which is really a collection of several texts by different authors) and Aryabhata's *Aryabhatiya*, which studied the relationship between half the chord and half the angle subtended by the chord. The Sanskrit term for the half-chord, 'jiva', became, over the years, the word 'sine' [16, 68], so the sine of an angle is the length of the half-chord (note this is a length; sine as a ratio is a more modern convention). If we create a second chord AB' perpendicular to AB in Figure 1.5, there is another angle at the center which is complementary to the first; its sine is then complementary to the sine of the first, and this 'complementary sine' became 'cosine'.

By the mid-9th century all the trigonometric quantities like tangent and secant that we use today appeared in Islamic texts, but what Islamic

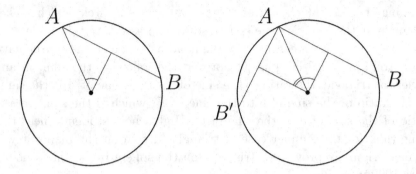

Figure 1.5: The sine and cosine of an angle as the lengths of half-chords in circles, according to the *Siddhantas* and the *Aryabhatiya*.

scholars were really interested in was *spherical* trigonometry, the reason being at least partly because it was required to face Mecca when praying and so understanding directions on the spherical Earth was crucial. The preliminary ideas were already in the Greek texts: a plane passing through the center of a sphere will intersect the sphere in a curve known as a 'great circle', for example the equator, and arcs of three great circles make a spherical triangle; see Figure 1.6.

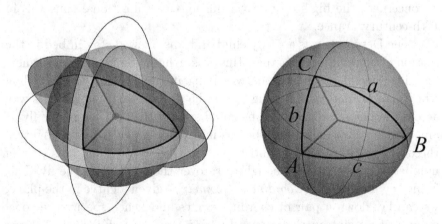

Figure 1.6: Three planes passing through the center of a sphere thus forming three great circles, and hence triangles, on the sphere.

If we label the vertices A, B, C, and the edge opposite each vertex a, b, c, then we can define an angle for all six: a vertex is where two great circles meet, so the angle at that vertex is the angle between the planes defining those great circles; an edge is an arc of a circle which makes an angle at the center of the sphere so we can associate this angle with the edge. In the 10th century the Baghdad astronomer and mathematician Abū'l-Wafā' al-Būzjānī [68] proved the following theorem: in any spherical triangle, the ratio of the sine of a to the sine of A is the same as the ratio of the sine of b to the sine of B, which is the same as the ratio of the sine of c to the sine of C. High-school students meet this 'sine rule' for the simpler case of triangles drawn on the plane; it was in fact originally proved for triangles on the sphere by Islamic scholars over 1000 years ago.

The big wheel of life turns, and by the 15th century Baghdad was no longer the scientific center it had been for several hundred years; instead wealthy traders and patrons from Europe began importing mathematical texts, originally written in Greek and Sanskrit, then Arabic, and then translated into Latin and European languages. With Europe in the ascendant economically, a new phase of the development of Mathematics began, with the initial emphasis on counting, problem solving, equations and then Algebra, as we will see in Part II. As far as Geometry is concerned, the big step forward during this transitional time was in 17th-century France.

René Descartes was a sickly child and was allowed stay in bed in the morning, thinking and writing; this was a habit he kept up his whole life. The legend has it that he was lying in bed looking at the ceiling which had beams criss-crossing it, and he observed a fly making its way across the beams. He realized he could specify the position of the fly at any moment by counting the beams in each direction - the fly might be three beams from the left and two from the bottom, and this pair of numbers $(3, 2)$ would be enough to recover the position of the fly. This is the key idea behind *Coordinate Geometry* - given a curve in the plane, we can lay down a pair of coordinate axes and hence specify the coordinates of a point on that curve. In 1637 Descartes published *Discours de la méthode pour bien conduire sa raison et chercher la vérité dans les sciences*, considered to be one of the founding moments in modern Western philosophy. Descartes promoted the sceptical approach: how do we know what we see is real? How do we know the world as we perceive it truly exists, and we are not just brains floating in a vat somewhere, being fed stimulation to make us perceive a certain reality? Descartes concluded the only thing a doubting mind can completely accept is its own existence, which he summed up in his famous phrase *cogito ergo sum*, I think therefore I am.

Discours de la méthode also contained three appendices, one of which was called *La géométrie* [31], in which he introduced his notion of a coordinate system which is called *Cartesian* in his honour. This is perhaps a little generous; what is actually in *La géométrie* is quite far from what we would consider coordinate geometry today: he does not always use perpendicular axes but opts for oblique instead, and there is no mention of formulae for distance or slope and so on. What's more, Descartes' contemporary and fellow countryman, Pierre de Fermat , had a similar idea at around the same time which was not published till much later, and, going back even further, Apollonius in his *Conics* also introduced a coordinate system based on the diameters described previously. Nonetheless *La géométrie* marked a turning point, and the use of coordinates coupled with the new-fangled Algebra provided mathematicians with a powerful new tool for understanding geometrical problems and then using this as a springboard to consider physical phenomena, as will see in subsequent chapters. We could say coordinate geometry is akin to a nursery, where some of the *enfants terrible* of mathematics were born and raised before growing strange and striking out on their own: Analytic Geometry, Calculus, Linear Algebra, all came out from under the apron of coordinate geometry. As for Descartes, he probably does not get the credit among mathematicians he deserves: he (co-)invented coordinate geometry, he introduced some of the founding ideas of Calculus (Chapter 9), he anticipated Euler's polyhedron formula by 100 years (Chapter 13), and he also seems to have anticipated the Matrix movies by several hundred years. Perhaps the reason is that he was by all accounts a pretty nasty man. He was vain and arrogant, and dismissive of other mathematicians; he was particularly harsh to Fermat whose work Descartes described as "merde" (pardon my French). In 1649 Descartes was invited to Sweden as tutor to Queen Christina who insisted on their sessions starting at 5am during one of the coldest winters in decades; René was dead within a couple of months.

How is coordinate geometry so different to the geometry that proceeded it, the geometry of Euclid and Apollonius? The difference is that coordinate geometry is more formulaic: given the coordinates of points such as the vertices of a triangle, we can calculate things such as lengths and areas directly. For example if two points have coordinates (A_1, A_2) and (B_1, B_2), then the midpoint of the segment joining them has coordinates $(\frac{A_1+B_1}{2}, \frac{A_2+B_2}{2})$, which is just the averages of the respective coordinates. If you have a line, then you can define a notion of 'slope': you just take two points on the line, say (A_1, A_2) and (B_1, B_2), and the slope is $B_2 - A_2$ divided by $B_1 - A_1$. Now we can easily determine if two lines are parallel: we just check do they have the same slope.

Consider the following theorem: *if you take any quadrilateral (for example the one on the left) and join the midpoints of successive sides, you always get a parallelogram.* How would you prove this in the manner of Euclid? Maybe join opposite vertices, drop perpendiculars, look for similar triangles? Would you need to worry about the different types of quadrilateral, ◁ or ▷ as opposed to ✕?

A coordinate geometry approach would go as follows: we choose a pair of axes, and then suppose the vertex A has coordinates (A_1, A_2), vertex B has coordinates (B_1, B_2) and so on.

The line segment joining A to B has midpoint with coordinates $(\frac{A_1+B_1}{2}, \frac{A_2+B_2}{2})$, and we label this point α (Descartes and his contemporaries would have been reading the Greek texts for inspiration; it seemed natural then to use Greek letters as symbols and mathematicians have been doing so ever since. On my first day in university I was given a print-out of the Greek alphabet and told to get used to it; how exotic!). The line segment joining B to C has midpoint

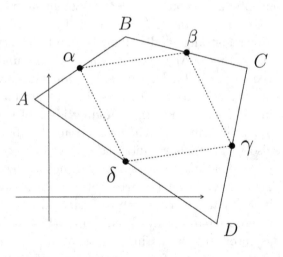

with coordinates $(\frac{B_1+C_1}{2}, \frac{B_2+C_2}{2})$, and we call this point β. Therefore the line joining α to β has slope $(C_2 - A_2)/(C_1 - A_1)$. In the same way we can find the coordinates of the midpoint of the segment joining C to D, we call this point γ, and the midpoint of the segment DA, we call this point δ. Now the line joining γ to δ *also* has slope $(C_2 - A_2)/(C_1 - A_1)$, so the segments $\alpha\beta$ and $\gamma\delta$ are parallel. A similar calculation would show that the segments $\alpha\delta$ and $\beta\gamma$ are also parallel, therefore $\alpha\beta\gamma\delta$ is a parallelogram.

What would Euclid have made of that? It is certainly lacking some of the elegance of the *Elements*, and feels a bit like a sledgehammer. Still, there is no denying its power but also its directness: there was no need to get creative or consider special cases, if we want to know something we simply calculate it. This is at least partly why most 14-year-olds can calculate the slope of a line but have never heard of Euclid.

You may notice I have been largely avoiding writing equations or using algebraic notation, and that is because up until the 16th century pretty much all mathematics was expressed in words and pictures; however now that our story has caught up with this time we will begin to use more modern expressions and notation. We are reaching the point in history where Geometry is getting too big to talk about all at once; it has begun to split into different camps with their own goals and methods, and that is why we will now look more deeply at some of those distinct branches.

CHAPTER 2

Non-Euclidean Geometry

L OOK BACK at the postulates of Euclid's *Elements*, and you will see the fifth one seems to stand out from the others. I have given here Playfair's wording of 1795 (Euclid's original wording was much more involved), but nonetheless the fifth postulate is harder to accept at face value and seems like it should instead follow from the others, as a theorem rather than a postulate. Already in the 5th century CE Proclus [41] says *this ought to be struck from the postulates altogether. For it is a theorem - one that invites many questions* and Omar al-Khayyāmī (11th century CE [41]) says *the book [Euclid's] Elements of Geometry which is the origin of all mathematics ... but there are doubtful matters, among them, the greatest one which has never been proved [5th postulate] ... I have seen many books which have objected to this idea, among the earlier ones Heron and Autolycus, and the later ones al-Khazen, al-Sheni, al-Neyrizi, etc. None has given a proof.* In the modern era several prominent mathematicians tried and failed to prove 'Euclid V' as it came to be known, in fact a Ph.D. thesis of 1763 was submitted finding the flaws in 28 different failed attempts [51]. Mathematicians were getting discouraged; d'Alembert called it "the scandal of geometry". What was needed was someone to cut the Gordian knot; it turned out three people wielded the knife at the same time, unbeknownst to one another.

In the 1820s the Hungarian mathematician Farkas Bolyai wrote to his son János, also a mathematician, with this warning: *You must not attempt this approach to parallels. I know this way to its very end. I have traversed this bottomless night, which extinguished all light and joy in my life ... I accomplished monstrous, enormous labors ... I turned back*

DOI: 10.1201/9781003455592-2

unconsoled, pitying myself and all mankind. Strong words! But when has a strong-headed son ever listened to his father? János had the vision to step beyond his father and all the other great names who had tried and failed before, and boldly realized: the reason no-one could prove Euclid V is that it is not necessarily true! It is perfectly logical and consistent to create a geometry where the first four postulates of Euclid hold, but the fifth does not. Note we say 'a' geometry here, for what results is a geometrical structure completely different to that of Euclid which had been the final word in geometry for the previous 2000 years. János writes in 1823 *I have discovered such wonderful things that I was amazed ... when you, my dear Father, see them, you will understand; at present I can say nothing except this: that out of nothing I have created a strange new universe.* This new universe had a bizarre distorted structure, like walking on a giant waterbed: not just one parallel line through a point but infinitely many; triangles that bend and deform as they are scaled up and down; a world where squares and rectangle cannot exist. Farkas was publishing a book, the *Tentamen*, surveying attempts at proving Euclid V, and he included his son's discoveries as an appendix; in 1831 he sent a copy to his friend and fellow mathematician, Carl Friedrich Gauss.

Gauss is widely recognized as one of the greatest mathematicians ever. We will say more about him as we go on, but for now let me mention Gauss's motto: *pauca sed matura*, few but ripe. Gauss would brood for many decades over his work, waiting until he felt it had reached a satisfactory level of completeness before publishing (hard for the modern academic to understand in the current 'publish or perish' culture). He would be visited by aspiring mathematicians to tell him of their results, only for Gauss to reach into a drawer and pull out some dusty papers where he had already arrived at the same results years before. His notes and 'partial' works were published incrementally after his death, and it has been said that if he had published his results during his lifetime then mathematics would be 50 years ahead of where it is today. After reading János's work in the appendix to the *Tentamen*, Gauss replied to Farkas: *If I begin with the statement that I dare not praise such a work, you will of course be startled for a moment: but I cannot do otherwise; to praise it would amount to praising myself; for the entire content of the work ... coincide[s] almost exactly with my own meditations which have occupied my mind for from thirty to thirty-five years.* Gauss refused to publicize the work, and János was not just devastated; our Icarus, burned, raged like the sea. A man of fiery temperament who had won more than 13 duels [4], Gauss was perhaps lucky he lived hundreds of miles away. Gauss was also cautious that he didn't want to get drawn into a public row over the foundations of geometry, explicitly saying he only wanted his results in this direction to be published after his death. He feared the scientific community at large would not be able to stomach the suggestion that Euclidean geometry was 'wrong'; Kant (1781) had said that *the concept of [Euclidean] space is by no means of empirical origin, but it is an inevitable necessity of thought* and few wanted to argue with him. Indeed János's father continued to publish 'proofs' of Euclid V, even though his own son had proven that the fifth postulate is not a necessary consequence of the previous four. Long-held beliefs are hard to overcome, and mathematicians are people after all.

Independently of this, the Russian mathematician Nikolai Lobachevsky had published in 1829 his own description of a geometry [68] that does not assume the validity of Euclid V, very like János Bolyai's, but it was little appreciated. In fact, a Russian literary journal scolded Lobachevsky for *the insolence and shamelessness of false inventions*; perhaps Gauss had been right to be wary. The new type of geometry described by these three men was originally called 'Lobachevskian', but in time came to be known as 'non-Euclidean'. The basic premise is this: the geometry of Euclid, 'Euclidean' geometry, takes as its starting point the first four axioms of the *Elements* plus the fifth: given a line l and a point P not on l there exists one line through P parallel to l. In non-Euclidean geometry we instead take the first four axioms of the *Elements* plus the *opposite* of the fifth. Now there are two choices: either there are no lines through P parallel to l, or there are infinitely many. We refer to these as 'elliptic' and 'hyperbolic' geometry respectively, echoing Apollonius's terminology for conic sections. We focus here on the two-dimensional versions, called the elliptic plane and hyperbolic plane, just as a sheet of paper might represent the Euclidean plane, but everything extends to higher dimensions so it makes sense to talk of 'Euclidean space' and 'hyperbolic space'. It is possible to treat these three geometries entirely formally, by which I mean treating 'points' and 'lines' as simply terms which satisfy certain relations and proceed deductively; it *can* be done this way, but it's hard work [51]. Instead what we need is a *model*, some way of representing the points and lines of the geometry. We have in fact already met a model of elliptic geometry: if we take the 'points' to be points on a sphere, and 'lines' to be the great circles of the sphere, then there are no parallel lines since all lines intersect (see Figure 1.6).

The elliptic geometry of the sphere will pop up again this book, but in this chapter we will instead dwell on a famous model of hyperbolic geometry: the *Poincaré disc* model of the hyperbolic plane (Poincaré 1882 and 1902). We define it like this: the 'points' of the hyperbolic plane are all the points inside (but not on) a circle, let's call it C (I will brush over the implications of the choice of radius for C). The 'lines' are the arcs of circles which intersect C at right angles (we might call them 'hyperbolic lines', or h-lines). That such hyperbolic lines must exist is a Classical result:

Construction 5 *Given a circle C and two points P and Q on C (but not diametrically opposite), construct the circle which intersects C at right angles at P and Q.*

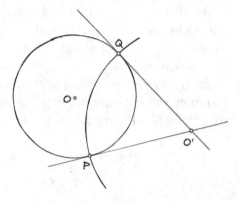

Construct the tangents to C at P and Q (Construction 3); continue them until they meet at the point O'; place the spike of the compass at O' and the pencil at P or Q to sweep out the desired circle. □

Now given a point in the hyperbolic plane (i.e. a point inside the circle) we can draw several hyperbolic lines through it:

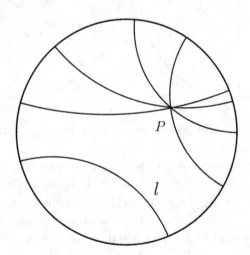

Observe the hyperbolic line l - there are several hyperbolic lines through P which are parallel to l in the sense that they never intersect it; thus Euclid V does not hold.

It is tempting to see the center of the circle above as the 'center' of the hyperbolic plane, and the h-lines as short segments, but this is not the case, just like if you took a sheet of paper and put a point in the middle you wouldn't think of that point as the 'center' of the Euclidean plane. Instead imagine flying in an airplane over the hyperbolic plane, and in the floor there is a viewer with a fish eye lens which lets you see all the way into the distance in all directions; this is what we are seeing in the picture above. As you fly over the plane the features of the landscape pass beneath you, the h-lines coming and going, and the 'center' of your view is nothing more than the point that happens to be beneath you. What's more, distances are distorted in the hyperbolic plane: if you were travelling along a h-line and at *equally spaced intervals* you drew the (unique) h-line perpendicular to your path, you might see something like Figure 2.1 on the left.

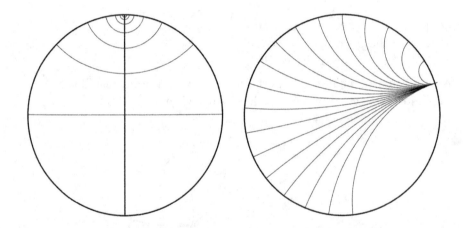

Figure 2.1: Hyperbolic lines in the hyperbolic plane.

The h-lines only seem to get closer since we are looking at them with our 'Euclidean' eyes; in the hyperbolic world these h-lines are all equally spaced and there is always room to draw another. As such you never reach the circle C, it is 'at infinity'. This means that when we draw a h-line in the hyperbolic plane we are drawing the *entire line*, its whole infinite length; in the Euclidean plane we can only draw a finite portion of a (Euclidean) line.

Although points on C are not in the hyperbolic plane it is convenient to give them a name: 'ideal' points, for example in Figure 2.1 on the right are several h-lines which all meet at the same ideal point. In Construction

5 above we saw how to draw a h-line in this model of the hyperbolic plane if you begin with two ideal points, but to be a valid model of non-Euclidean geometry we need to satisfy postulates I-IV, the first of which is 'a line segment can be drawn from any point to any other point'. In our context we can phrase this as: given two points inside the circle C, can we draw a unique arc of a circle which intersects C at right angles and passes through the two given points? This is again a straightforward compass and straight-edge construction; we won't give it here but the key step follows from a theorem first given in Euclid's *Elements* (III-36).

A hyperbolic triangle is simply the polygon formed by three hyperbolic lines, for example on the left below:

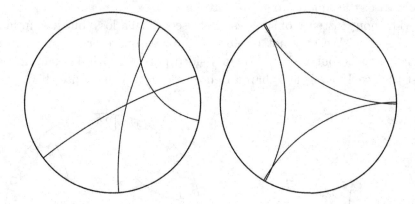

You might notice the vertices seem somewhat 'pinched'; a more extreme example where the vertices are further apart is on the right. Though lines in the hyperbolic plane appear distorted to us the angles are not, and so it is clear that the three angles of these triangles most certainly do *not* add up to two right angles (also known as 180°, but we will find it more convenient to write this as 'π radians'). One of the central results of hyperbolic geometry is the following:

$$\left(\begin{array}{c} \text{sum of the angles in} \\ \text{a hyperbolic triangle} \end{array} \right) = \pi - \left(\begin{array}{c} \text{area of the} \\ \text{hyperbolic triangle} \end{array} \right). \qquad (2.1)$$

So the three angles of a hyperbolic triangle always add up to *less than* two right angles or π radians. The angle sum is almost π for small triangles (on the left above), but it gets smaller and smaller as the triangles get bigger (on the right above). This theorem was already known to Gauss in 1794, when he was only 17 years old! Two immediate consequences are: since an angle sum cannot be negative, the right hand side of (2.1) must be positive and so the area of a triangle is at most π, even if the vertices are very far apart, but also there are *no similar triangles* in the hyperbolic plane, since when you try to scale up a triangle you necessarily change the area and hence the angles. The result extends to 4-sided polygons:

$$\begin{pmatrix} \text{sum of the angles in} \\ \text{a 4-sided polygon} \end{pmatrix} = 2\pi - \begin{pmatrix} \text{area of the} \\ \text{4-sided polygon} \end{pmatrix}. \qquad (2.2)$$

We see now why squares, with four right angles, do not exist in the hyperbolic plane: if all the angles at the vertices were $\pi/2$ radians, then the area of the square would need to be zero.

The non-existence of squares might seem like a loss, but it is in fact the opposite! We may not be able to draw a square or rectangle with 4 right angles, but we can draw a 4-sided polygon with 4 equal angles, just that each one is less than a right angle. How can we use this?

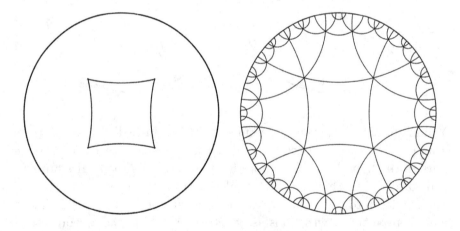

Well if you are ever in a hardware store you will notice wall tiles tend to be square. This is because you can fit squares together without any gaps or overlaps; we would say that squares 'tile' the Euclidean plane. A square is 'regular', in that all the sides and angles are equal, and there are only three regular polygons that tile the Euclidean plane: squares, triangles and hexagons. In each case the angle at the vertex is just right so you can fit 4, 6 or 3 polygons around a vertex without any gaps or overlaps. In the hyperbolic plane however, we could have a 4-sided regular polygon, like the ones in the images above, only now the angle at

Figure 2.2: Two tilings of the hyperbolic plane [2]: above, six 'squares' meet at a vertex, and below, eight 'octagons' meet at a vertex.

each vertex is less than $\pi/2$, which means you could potentially fit more than 4 around a vertex, again with no gaps or overlaps. Suppose the angle at the vertex was $\pi/3$, now you could perfectly fit 6 'h-squares' around a vertex. We even know from (2.2) exactly how big such a h-square would need to be: $2\pi/3$. As such we can tile the hyperbolic plane with regular 4-sided polygons, only now with 6 meeting at each vertex.

The important thing to note about this regular tiling is: all the polygons are the same size, since they all have the same angles at their vertices ($\pi/3$ radians), and therefore they all have the same area by (2.2). If we were flying in our airplane over this tiling of the hyperbolic plane, we would see the polygons passing below us, each one an identical copy of the last. The images in Figure 2.2 makes this more clear by supposing our viewpoint is not directly over a vertex. Actually early versions of these hyperbolic tilings were already known before Poincaré, to Riemann in 1858 and Schwarz in 1872, and Gauss was up to his old tricks: drawings by Gauss just like these were found years after his death, tucked away in a dusty drawer [109].

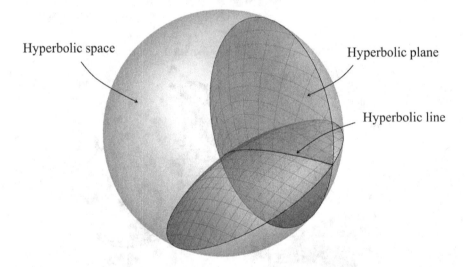

Figure 2.3: The Poincaré 'ball' model of hyperbolic space. Note the intersection of two hyperbolic planes is a hyperbolic line, as expected.

Finally we can easily extend this model of the hyperbolic plane to a model of hyperbolic space: we take all the points inside a sphere, with hyperbolic planes as the portions of spheres that intersect this sphere at right angles; see Figure 2.3.

You might be saying: all this is fine, but hyperbolic geometry isn't 'true' as such, is it? To which the slightly smarmy answer would be: is Euclidean geometry 'true'? Can you draw a perfect circle? Can you draw two infinitely long parallel lines? Both Euclidean and non-Euclidean geometries are formal systems, with objects and relations. One is not more 'true' than the other, and both Euclidean and non-Euclidean geometry are perfectly consistent, there are no internal contradictions. Nonetheless you might feel that Euclidean geometry is closer to reality than non-Euclidean, and certainly in the everyday you would be right: good luck trying to tile your kitchen using hyperbolic 'squares'. But the 19th century saw a profound shift in thinking within mathematics, a line had been crossed in the drive toward further and further abstraction. Yes Euclidean geometry may deal with ideal notions of lines and circles, perfect and abstract, but they are still perfect and abstract representations of 'real' objects. The far-reaching developments of Calculus and Algebra that we will see in subsequent chapters were still rooted in the natural sciences, they were still ultimately motivated by physical applications. But non-Euclidean geometry was a revolution: it is valid and of interest even though it claims no connection to the real world and the natural sciences. Some argue that this is the beginning of what can be called 'pure' mathematics, and there was considerable push-back at the time as Gauss foresaw: there was a genuine worry that *mathematicians had climbed so high up their mountain of abstraction that they could no longer find their way back to reality* [4]. But those critics didn't learn their lessons from history: mathematics is a broad church, and there is room for the pure and the applied. But it is more than that: the pure and the applied are not mutually exclusive, rather they intermingle and support one another. Non-Euclidean geometry is a good example of that: while Euclidean geometry is appropriate for tiling walls and building bridges, perhaps our universe is non-Euclidean on the larger, astronomical scale? Einstein said *to this [non-Euclidean] interpretation of geometry I attach great importance, for should I not have been acquainted with it, I never would have been able to develop the theory of relativity* [51]. To understand the geometry of the universe, and to develop an overarching framework that includes all of the geometries discussed in this chapter and more, we need to contemplate curvature.

CHAPTER 3

Curves, Surfaces, Manifolds

H OW CURVED is a curve? Is the cylinder curved, like a sphere? Since the Earth turned out to be curved, could the universe be curved too? The desire to understand curves and surfaces was one of the drivers behind the development of Calculus, and this interplay between calculus and geometry grew into its own branch of mathematics known as Differential Geometry; dealing with everything from doodles on a page to the fatal attraction of black holes, the recurring theme is curvature and how it dictates geometry, as we will see. While we hold off on the development of Calculus till Part III of this book, we will introduce here some basic notions such as functions and increments, while drawing a veil over the technicalities as much as possible. We will begin as always with the atom of Mathematics, the silver thread that runs through this book: curves in the plane.

Consider the two circles below:

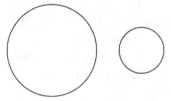

One is big and one is small. Suppose each was a road that you were driving your car along; as you drive along the circle on the left you

DOI: 10.1201/9781003455592-3

would only need to turn your steering wheel a little to the side, whereas with the circle on the right you would need to turn it much further. The circle with the smaller radius is *more curved*, and this suggests the curvature, k, of a circle with radius r should be

$$k = \frac{1}{r}.$$

This description of the curvature of a circle was first put forward by Nicole Oresme in 14th-century France, one of the few bright lights in medieval Europe [100]. If we imagine the radius getting larger and larger then the curvature will get smaller and smaller; if we think of a straight line as a circle with 'infinite radius' then the curvature of a straight line is zero, as we would expect.

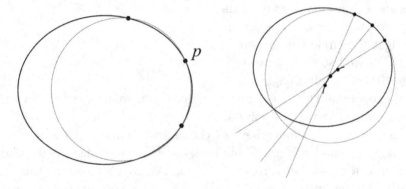

Figure 3.1: On the left: the circle through three points on an ellipse; on the right: the centers of the osculating circles to a curve lie on the evolute.

For a curve in general, let's say the ellipse on the left of Figure 3.1, we can define the curvature at an arbitrary point p in the following way: taking two points to either side of p, we know from Chapter 1 Construction 4 that we can construct a circle through these three points. As the two neighboring points get closer and closer to p the circle seems to 'fit' the curve at p better and better; the circle of best fit was called the 'osculating' or kissing circle by Leibniz in the 1670s. Descartes in *La géométrie* considered the osculating circle [16], but he used it to construct the tangent line at p: the radius of the osculating circle at p is perpendicular (or 'normal') to the curve there and thus rotating it gives the tangent. Huygens in the 1660s was interested in osculating circles but for yet another reason: if you take successive points on the curve (see Figure 3.1 on the right) and trace out the centers of the osculating circles you get another curve which has special properties he used in the design of pendulum clocks; Huygens called this second curve the 'evolute' [137] (see also Figure 11.3) but it seems Apollonius anticipated evolutes all the way back in his *Conics* [55].

Now we can define the curvature of a curve at a point: it is simply the curvature of the osculating
circle at that point. On the right
we show an ellipse with several of
its osculating circles; we can see
at the point on the top the osculating circle has the largest radius and so the curvature is smallest there, whereas at the point on
the right the circle has the smallest radius and so the curvature is
largest there (again imagine driv-

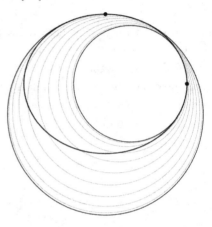

ing a car around the ellipse and how much you would need to turn the steering wheel at various points).

But we want to move beyond the classical curves of the ancients, beyond circles and ellipses. Think back to Descartes lying in bed looking up at the fly crawling across his ceiling. At any moment in time the position of the fly is given by a pair of x and y coordinates, but since the fly is moving those coordinates will change with time t; we say the coordinates are *functions* of time, and write $x(t)$ and $y(t)$. The notion of function developed slowly over the centuries and we will see it again as we go on, but for now we can think of functions like a machine with an input and an output: you feed in a value for t, it spits out a value for x or y. Here t is called a 'parameter', and this way of expressing a curve is called the 'parametric form' or 'parameterization'; time may be a natural choice as parameter, but there are other choices as we will see.

Using osculating circles, Newton in 1671 wrote down a formula for the curvature of a curve in terms of the parameterization, saying *there are few problems concerning curves more elegant than this* [53]. In fact this was one of his first demonstrations of the usefulness of his *fluxions* [68] that we will see in Chapter 9, because the formula applied to *any* curve, whereas previously an elaborate bespoke construction was needed for each specific curve. We note that Newton preferred this parameterized form as it is a more dynamic way of describing a curve: where the ancients had seen circles and ellipses as static and fixed, Newton instead thought of a curve as the path traced out by an object, such as a fly or a planet, as time goes by. Let's do an example.

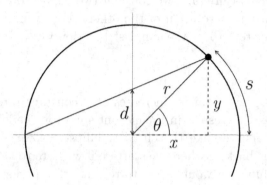

The standard parameterization for a circle of radius r goes like this: we draw a pair of coordinate axes with origin at the center of the circle. The x coordinate of the black dot is the length of the horizontal leg of the right angled triangle shown dashed, and if θ is the angle in the diagram then $x = r\cos(\theta)$. Similarly the y coordinate is the length of the vertical leg so $y = r\sin(\theta)$; hence a parameterization of a circle of radius r could be

$$x = r\cos(\theta), \qquad y = r\sin(\theta). \qquad (3.1)$$

Here x and y are functions of θ, and feeding into these functions different values for θ will give you different points on the circle; allowing θ to vary continuously will give a sense of the black dot moving around the circle.

There are many ways to parameterize a curve because there are many ways to identify a point on a curve; for example look at the length d in the diagram. As the black dot moves around the circle then d will vary, and in the same way if you were to specify a value for d this will identify a point on the circle; as such we could write the x and y coordinates as functions of d and this would be another parameterization of the circle (it is likely Diophantus anticipated this approach to distinguishing points on a circle in the 3rd century CE [109]). But perhaps the most important way to parameterize a curve is to choose some point on it and then measure the length along the curve from that point; this is known as the 'arc-length' parameterization, usually denoted s. Whatever parameterization you use, you will always find the curvature at a particular point to be the same.

The curvature of a curve in the plane turns out to be *the* defining characteristic of the curve [35], in the sense that if you have a plane curve then you can calculate its curvature, but this works in reverse: if you specify a function first, there is a unique curve which has that function as its curvature [57] (we consider two curves to be 'the same' if one is a translation or rotation of the other). For example, suppose we want the curvature k to be the simplest function we can think of:

$$k = s,$$

in other words, the curvature increases at a constant rate as we move along the curve; this results in the elegant spiral on the left below.

For 'closed' curves like the circle or ellipse, as we go all the way around we come back to where we started; if we were to keep going then the curvature function would start to repeat, it is 'periodic'. All closed curves have periodic curvature functions, but this does not necessarily go the other way: just because the curvature is periodic does not mean the curve is closed [8] (in the middle and on the right below are the curves with curvature $k = \sin(s) + 3\cos(s)$ and $k = \frac{1}{5} + 2\sin(s) - \cos(2s)$ respectively, but only one is closed).

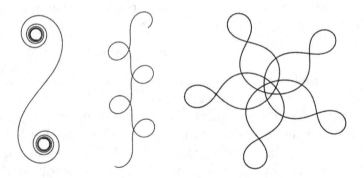

Now we have an idea of curves, what about surfaces?

Enter Euler. Leonhard Euler was like the 'anti-Gauss': whereas Gauss's motto was *pauca sed matura*, Euler's might have been *copiosus et perspicuus*. Euler wrote and published prolifically on pretty much every aspect of mathematics, the old rub being he could dash off a paper while waiting for his dinner to cook. Though born to humble circumstances,

Euler took positions in the St. Petersburg and Berlin Academies, and in his career he produced over 800 books, papers and notes (and 13 children, although his wife probably deserves the credit for that). For example his writings made up half the pages published by the St. Petersburg Academy of Sciences from 1729 until over 50 years *after* his death [109]. He went blind in one eye in 1738, and then completely blind in 1771, but his output just increased in response; his huge contribution is honoured by the many theorems, equations and methods named after him and we will see several in this book. Laplace had this advice for his fellow mathematicians: *read Euler, read Euler, he is the master of us all.*

In 1760 Euler published *Recherches sur la courbure des surfaces* [39] in which he described a way of measuring the curvature of a surface at a certain point by thinking about the curvatures of the curves lying in the surface and passing through that point; the problem is there are very many curves which may pass through the same point, but also that our description of curvature above was only for *plane* curves. Euler solved both of these problems in the following way: for the

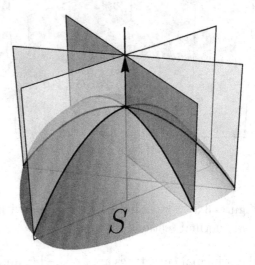

Figure 3.2: A portion of a surface S with the normal line and some normal sections shown.

surface S in Figure 3.2, take the line normal (or perpendicular) to the surface at a point, and then consider all the planes which contain that normal line. Each of those planes will intersect the surface thereby giving a curve (called the 'normal section') which lies in the surface but also in a plane; thus we can think about the curvature of those curves.

Now imagine rotating the planes about that normal line; as we do the normal sections will vary and hence their curvature will vary. Euler showed that there will be a maximum and minimum value for those curvatures, and in fact the planes that give those curves with the maximum and minimum curvatures will be at right angles to one another; he called these the 'principal curvatures', which we will label k_1 and k_2. We can put a sign on the principal curvature: we give a direction to the normal line (indicated with an arrow in Figure 3.2), and if the normal section is curving in the direction of the normal line we say the curvature is positive, but if it is curving in the opposite direction we say it is negative. In Figure 3.2 the principal curvatures are both negative (we could say the normal sections are curving 'away from the normal line').

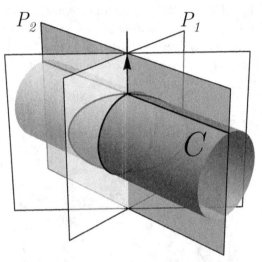

Consider for example the infinite cylinder of radius r. Some planes containing the normal line are shown in Figure 3.3, and we can see that the section with the largest curvature is when the plane is perpendicular to the axis of the cylinder, labeled P_1, in which case the normal section is a circle of radius r. The section with the smallest curvature is when the plane contains the axis of the cylinder, labeled P_2, in which case the normal section is a straight line. Thus the principal curvatures are $k_1 = -1/r$ and $k_2 = 0$ (the minus is because the circular section is curving *away* from the surface normal).

Figure 3.3: The infinite cylinder C with some normal sections shown.

Actually I am just following Euler himself here: in *Recherches* he made some definitions and did some calculations and then followed this up with several examples, such as the cylinder (see Figure 3.4) and the cone, even going so far as to derive the formulae for the principal curvatures for an ellipsoid (the ellipsoid is to the sphere as the ellipse is to the circle; see Figure 11.8). Euler was in fact quite renowned for taking pains to explain his results using examples and straightforward language, not a trait shared by all scientists unfortunately.

There is an ambiguity however: think back to the spherical trigonometry from Chapter 1 and Figure 1.6; *all* the normal sections of a sphere of radius r (see Figure 3.5) are great circles which must also have radius r and therefore curvature $1/r$, so the principal curvatures are equal to one another, but what sign should they have?

❀ *133* ❀

une autre fection quelconque, qui eft inclinée à la principale de l'angle $= \varphi$,

il faut prendre $s = \dfrac{- a \, \text{tang} \, \mathcal{D}}{V(aa - yy)}$: & alors le rayon ofculateur fera:

$$- \frac{(1 + qq + ss)}{- aa : (aa - yy)^{\frac{3}{2}}} \cdot \frac{a}{V(aa - yy)} = \frac{1 + qq + ss}{a}(aa - y^2),$$

qui fe réduit à cette forme: $a(1 + \text{tang} \, \varphi^2) = \dfrac{a}{\text{cof} \, \varphi^2}$; d'où

l'on voit que pour la fection principale le rayon ofculateur eft $= a$, à caufe de $\varphi = 0$, & pour la fection qui y eft perpendiculaire & paffe par l'axe du cylindre, il devient infini: ce qui marque que la fection eft une ligne droite.

Figure 3.4: Excerpt from Euler's *Recherches sur la courbure des surfaces* [39]. Note how Euler refers to *le rayon osculateur*, the osculating ray, which is the radius of the osculating circle.

If I take an *outward* pointing normal, then the normal sections are curving away from it and so the principal curvatures will both be negative, but if I choose an *inward* pointing normal then the normal sections are curving toward it and so the principal curvatures will be positive: the sign of the principal curvatures depends on your choice of direction for the normal, which is a choice we make rather than a property of the surface (indeed for some surfaces 'inward' and 'outward' don't even make sense, as we will see in Chapter 16).

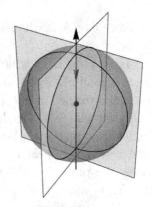

Figure 3.5: Normal sections on a sphere.

Enter Gauss. In the 1820s Gauss was occupied with a survey of the Kingdom of Hannover, and he thought deeply about the lines and triangles he was mapping on the curved surface of the Earth. He considered the following question: what is the *intrinsic* geometry of a surface? By which we mean, imagine a little ant crawling across the surface, only aware of the two dimensions of the surface itself and completely unaware of any external space; what is the geometry that ant would perceive? What would lines, circles and triangles be to the ant? Suppose it was the Euclid of the Ant World, what theorems would it propose? Gauss worked backwards from thinking about triangles on curved surfaces to the curvature of the surface itself and hence to the lengths of curves in the surface, publishing in 1827 the riper-than-ripe *Disquisitiones generales circa superficies curvas* [47]. In it, and in pretty much every Differential Geometry book since, Gauss presented his results in the reverse order of how he discovered them, and we will do the same.

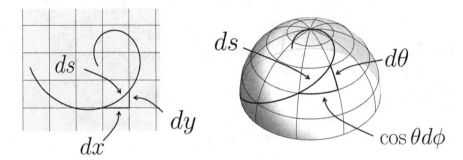

Figure 3.6: A small element of a curve in the plane and on a sphere.

If we remember what the word 'geometry' means literally, this suggests a good place to start would be measuring lengths of curves. How do we measure the length of a curve in a surface? Often the way to address questions like this is to drop down to a simpler context that you already understand well but phrase it in a new way that then generalizes, so we ask: how do we measure the length of a curve in the plane? Imagine we cover the plane in an x-y coordinate system, like on the left in Figure 3.6. To measure the length of the curve, we zoom in on a little element of the curve; zoom in far enough until it starts to look like a line. If we let ds denote the length of this little element of the curve, then the difference in the x-coordinates of the end-points is dx, and the difference in the y-coordinates is dy. Pythagoras theorem then says

$$ds^2 = dx^2 + dy^2.$$

To find the length of the whole curve, we add up all the lengths of these little elements.

But now suppose instead we have a curve on the sphere, like the one shown on the right in Figure 3.6 (we let the radius of the sphere be 1 to make things easier for now). Following the same approach, we first need a coordinate system such as the latitude-longitude coordinates, which we will label θ and ϕ. Now when we zoom in on a little element of the curve, its length ds will depend on the difference in the θ and ϕ coordinates of the endpoints, $d\theta$ and $d\phi$, but there is a complication: if you look at the coordinate system on the sphere, you will see the lines of longitude get closer and closer together as we approach the poles, even though the difference in the ϕ coordinate is the same for every pair of points on neighboring lines of longitude. To correct for this, the length of the horizontal leg of the triangle on the right of Figure 3.6 is not $d\phi$ but $\cos\theta\,d\phi$, so Pythagoras theorem now says

$$ds^2 = d\theta^2 + \cos^2\theta\,d\phi^2.$$

Expressions like this one and the one above are called the 'line element'.

If you have a general surface, with a pair of coordinates called u and v for example, then we will again have a line element and it will be some expression of the form

$$ds^2 = \left(\text{quadratic in } du \text{ and } dv.\right)$$

Gauss in the *Disquisitiones* decided to label the coefficients in this quadratic expression E, F, G and these symbols have stuck ever since: the general expression of a line element is

$$ds^2 = E\,du^2 + 2F\,du\,dv + G\,dv^2. \tag{3.2}$$

The reason we have the $du\,dv$ bit in the middle is that the coordinate lines may not be perpendicular to one another (like they are for Cartesian coordinates in the plane and latitude-longitude coordinates on the sphere), so we need to generalize Pythagoras theorem to the cosine rule. The line element is *the* fundamental object describing the intrinsic geometry of a surface, and our imaginary 2-dimensional ant would be able to write down (somehow) a line element for their world. Anything we can calculate from the line element, such as lengths, angles and areas, is part of the intrinsic geometry of the surface.

As for the curvature of the surface, let us return to Euler's principal curvatures described in the last section. We noted that the sign of the principal curvatures would change if we decided to let the normal point one way rather than the other, however if we look at the product of the principal curvatures,

$$K = k_1 k_2, \tag{3.3}$$

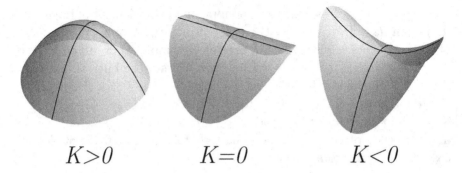

$K>0 \qquad K=0 \qquad K<0$

Figure 3.7: The neighborhood of a typical point in a surface looks like one of the three cases above, depending on the sign of the Gauss curvature.

then that does *not* change when we change the direction of the normal (if you replace k_1 and k_2 with $-k_1$ and $-k_2$, then K is still $k_1 k_2$). This is known as the *Gauss curvature*, and in fact in the *Disquisitiones* Gauss first described the curvature in terms of how the normal varied as we move around the surface, but then quickly pointed out how that description is the same as (3.3). This is important, because the *sign* of the Gauss curvature at a point turns out to be the key distinguishing feature (see Figure 3.7): if the Gauss curvature at a point is positive then the surface is like a 'bowl' near that point, whereas if K is negative then it is like a 'saddle'; if K is zero then the surface is 'flat', like a cylinder.

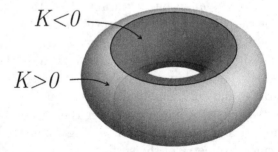

$K<0$

$K>0$

The Gauss curvature can vary from point to point; above is a 'torus', like the tube inside a bicycle wheel or the glaze on a ring donut. The inner region has negative curvature (the neighborhood of any point in that region looks like a saddle), whereas the outer part is positively curved; in fact we will see the torus has precisely as much positive curvature as negative (see the discussion around Figure 13.5).

ctiones indeterminatarum p, q, vnde pro elemento $\sqrt{}\ (\mathrm{d}x'^2 + \mathrm{d}y'^2 + \mathrm{d}z'^2)$ prodibit expressio talis

$$\sqrt{}\ (E'\mathrm{d}p^2 + 2F'\mathrm{d}p \cdot \mathrm{d}q + G'\mathrm{d}q^2)$$

denotantibus etiam E', F', G' functiones ipsarum p, q. At per ipsam notionem *explicationis* superficiei in superficiem patet, elementa in vtraque superficie correspondentia necessario aequalia esse, adeoque identice fieri

$$E = E', \ F = F', \ G = G'.$$

Formula itaque art. praec. sponte perducit ad egregium

THEOREMA. *Si superficies curua in quamcunque aliam superficiem explicatur, mensura curuaturae in singulis punctis inuariata manet.*

Figure 3.8: Excerpt from Gauss's *Disquisitiones* [47]. Note his use of the word *egregium*.

When Gauss went to derive a formula for the Gauss curvature, he discovered something remarkable: the Gauss curvature can be written entirely in terms of the coefficients of the line element. He was so impressed by this he referred to it as his *theorema egregium*, his wonderful theorem; see Figure 3.8. Gauss proved many theorems over the course of his career, sometimes returning to the same theorem to prove it several times as we will see in Chapter 6; what is so special about this theorem he called it wonderful? Look at it this way: remember our little ant crawling around their two dimensional world with no concept of an external ambient space. This ant could make measurements to derive a line element and from that they could calculate the Gauss curvature of their universe, and then from that infer the existence of some previously unknown external dimension!

But Gauss didn't stop there - this fruit was bursting with ripeness and the seeds were spilling out. What is the inner geometry of a curved surface like? How does it differ from that of the plane? Gauss took a basic theorem from Euclidean geometry: the three angles of a triangle add to 180° (or π radians), and investigated whether the same would hold if we were to draw a triangle on a surface. But to draw a triangle you need three straight edges; how do you draw straight lines on a curved surface? Gauss supposed that the sides of the triangle were *lineis breuissimis*, which translates as 'shortest lines'; we would now refer to these curves as 'geodesics' and we will see them again in Chapter 11. Gauss proved the following theorem [35]:

$$\left(\begin{array}{c} \text{sum of the angles in} \\ \text{a geodesic triangle} \end{array} \right) = \pi + \left(\begin{array}{c} \text{total Gauss curvature} \\ \text{inside the triangle} \end{array} \right), \quad (3.4)$$

in other words, it is the Gauss curvature of the surface that dictates the geometry. This theorem was extended by Bonnet in 1844 to include the case where the sides are not geodesics, and so it is known as the Gauss-Bonnet theorem (in fact this is the 'local' version of the theorem; we will see the global version in Chapter 13). If the Gauss curvature of the surface is a constant, let's say K_0, then the theorem goes

$$\left(\begin{array}{c} \text{sum of the angles in} \\ \text{a geodesic triangle} \end{array} \right) = \pi + K_0 \times \left(\begin{array}{c} \text{total area of} \\ \text{the triangle} \end{array} \right). \quad (3.5)$$

There are three cases: if K_0 is zero then the surface is flat, like the plane, and the three angles of a triangle add to π radians (or *two right angles*, as Thales put it). If K_0 is positive, then the surface is a sphere and the triangle is that formed by three great circles, like those studied by the Islamic scholars a millennium ago (see Figure 1.6); the sum of the three angles in such a triangle will be *greater than* π radians (Harriot first proved (3.5) for the sphere in 1603 [109]).

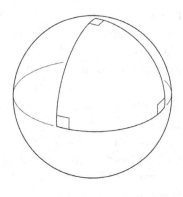

As an example, let's verify the theorem for the triangle on the right. There are three right angles, each one $\pi/2$ radians, so the left hand side of (3.5) is $3\pi/2$. The Gauss curvature of a sphere of radius r is $1/r^2$ (since, as we saw, the normal sections are all circles of radius r) and the area of the triangle is one eighth that of the sphere, $4\pi r^2/8$, so the right hand side of (3.5) is $\pi + \frac{1}{r^2} \times \frac{\pi r^2}{2}$ which is indeed equal to $3\pi/2$, as expected.

The final case, K_0 negative, we have in fact seen already but that will not be evident until the next section!

Enter Riemann. We have gone from Euclid, one of the most prolific mathematicians ever, to Gauss few-but-ripe, to Bernhard Riemann, who published only a handful of papers during his short life but nonetheless each one was groundbreaking in its depth and reach. Riemann was a

guren jede beliebige Lage gegeben werden. Die Massverhältnisse dieser Mannigfaltigkeiten hängen nur von dem Werthe des Krümmungsmasses ab, und in Bezug auf die analytische Darstellung mag bemerkt werden, dass, wenn man diesen Werth durch α bezeichnet, dem Ausdruck für das Linienelement die Form

$$\frac{1}{1+\frac{\alpha}{4}\,\Sigma\,x^2}\,\sqrt{\Sigma\,dx^2}$$

gegeben werden kann.

Figure 3.9: Excerpt from Riemann's *Über die Hypothesen* [97]; in fact this is the only equation in the work.

student of Gauss's in Göttingen, and when Riemann was put forward for his *habilitationsschrift* (which is like a conferring of professorship and includes presenting an inaugural lecture), Gauss suggested the title *Über die Hypothesen, welche der Geometrie zu Grunde liegen*, On the Hypotheses which lie at the Bases of Geometry. Riemann gave his inaugural lecture in 1854, and while there are many 'big bang' books in the history of Maths and Science, there are very few 'big bang' lectures; this is one. Colleagues were amazed when Gauss, possibly for the first time in his long career, expressed admiration for the work of another mathematician; praise from Caesar! You can see a YouTube video of the lecture here [88].

Riemann's approach was completely novel, revolutionary even. He boldly said we can take any space of any dimension, and simply assign a line element to it; behold! You have defined a new geometry. Gone is the need for any sort of ambient Euclidean space, gone is the Kantian constraint of three dimensions; a totally general and encompassing vision which includes the non-Euclidean geometries of Bolyai and Lobachevsky, and the Euclidean geometry of Euclid of course, as well as the surfaces of Euler and Gauss, and an infinitude of geometries still to be imagined, shorn free from the shackles of the physical world. All we needed, he said, was a way of measuring the distance between nearby points, a 'metric' provided by the line element, and from that he described how notions of curvature could be extended to any dimension. The word 'surface' is failing us now so Riemann used the term 'manifold' for these abstract spaces of arbitrary dimension, and Riemann's lecture was unusual in that it included only one equation: the line element for a manifold of constant curvature α (see Figure 3.9). Let's consider the case where the curvature is negative, and take $\alpha = -1$. In two dimensions the line element will then be

$$ds^2 = \frac{dx^2 + dy^2}{(1 - \frac{1}{4}(x^2 + y^2))^2}.$$

Look at the denominator: this vanishes when $x^2 + y^2 = 4$, so we must only include the points where $x^2 + y^2 < 4$, i.e. the points inside a circle of radius 2. Now looking for the *lineis breuissimis*, the geodesics, we find they are nothing more than the arcs of circles which intersect this circle at right angles; in other words, we have recovered the Poincaré disc model for hyperbolic geometry from the previous chapter, which we can now understand as simply the geometry of manifolds with constant negative Gauss curvature. What's more we see that Gauss's theorem for geodesic triangles, (3.5) with $K_0 = \alpha = -1$, is simply (2.1) again:

$$\begin{pmatrix} \text{sum of the angles in} \\ \text{a hyperbolic triangle} \end{pmatrix} = \pi - \begin{pmatrix} \text{area of the} \\ \text{hyperbolic triangle} \end{pmatrix}.$$

But it goes deeper than that: what about our three dimensional universe? Can we perform measurements of lengths and angles, and conclude that we are sitting in some sort of curved manifold, perhaps embedded in some higher dimensional reality? Ladies and gentlemen, we are the ants!

In fact Bessel attempted to measure the curvature of our universe using large scale astronomical observations; what's more, discrepancies between the predictions of Newtonian gravity and observations of planetary orbits (as we will see in Chapter 11), coupled with bizarre and unexplained phenomena in electromagnetism lead to unease in the scientific community at the end of the 19th century. Revolution was in the air. Then came the hand grenade.

Enter Einstein. Albert Einstein was a little know patent clerk in Bern, Switzerland when in 1905 he published four papers. Each one was significant, but the one that draws our attention here is *Zur Elektrodynamik bewegter Körper*, On the Electrodynamics of Moving Bodies [36]. This marked the beginning of Special Relativity, which proposes that our universe, rather than being 3-dimensional Euclidean space as had been assumed by pretty much every human being up to then, is in fact a 4-dimensional manifold called 'spacetime', defined in the manner of Riemann: we take the four coordinates x, y, z and t, and assign the line element

$$ds^2 = c^2 dt^2 - dx^2 - dy^2 - dz^2$$

where c is the speed of light. Spacetime is weird [115]: lines can be perpendicular to themselves and 'lengths' can be negative; you could fly at great speed to a distant galaxy and when you returned everyone you knew would be long dead, as if you had travelled into the future; if you were carrying a pole 10m long you could run at great speed into a room 5m deep and shut the door behind you; even the color of light changes at great speed, although that might not hold up in court ('sorry your honour I broke that red light, it appeared green since I was driving at 99% the speed of light. . . ').

Things get even stranger when you add gravity to the mix. Supposing the universe is completely empty except for a spherical mass m, spacetime has the line element [34]

$$ds^2 = \left(1 - \frac{2m}{r}\right) dt^2 - \left(1 - \frac{2m}{r}\right)^{-1} dr^2 - r^2(d\theta^2 + \cos^2 \theta d\phi^2), \quad (3.6)$$

the famous Schwarzschild solution. The standard way to visualize this is the analogy of the rubber sheet and the bowling ball, and we will draw pictures like that in Chapter 10, however the problem with that analogy is it requires the rubber sheet to be bending into some sort of external space, but the curvature of the Schwarzschild solution is purely *intrinsic*, there is no need for any external Euclidean space for spacetime to be sitting in. An alternative analogy is in the cartoon on the right: the spheres we see are really all equally spaced, but the mass of the Earth is curving spacetime around it and so distorting distances.

One last thing about (3.6): notice how some of the coefficients have terms like $\frac{2m}{r}$, which means when we calculate the curvature of spacetime it is infinite at the point $r = 0$. This 'singularity' from which not even light itself can escape is surrounded by a black hole in the Schwarzschild solution, but there are more elaborate models of spacetime where a rotating black hole drags spacetime around with it potentially opening a portal to another universe, or perhaps there is no protective black hole at all, a 'naked singularity' [129]; who knows what other cosmic monsters lurk in the depths of our curvature ocean?

CHAPTER 4

Fractal Geometry

A LL THE curves we saw previously had a common property: when you take a generic point on a curve and zoom in on it, and keep on zooming in more and more, that part of the curve will start looking like a straight line. Lines are one dimensional, and this is how we tell them apart from points and areas. There's even a nice hierarchy to the intuitive notion of dimension: if we consider a cube, the inside of the cube is a 3-dimensional volume, the faces of the cube are 2-dimensional areas, the edges of the cube are 1-dimensional lengths, and the vertices of the cube are 0-dimensional points. Surely this is rock solid? Who could argue with that? But as Benoit B. Mandelbrot said [79]: *clouds are not spheres, mountains are not cones, coastlines are not circles. . . Nature exhibits not just a higher degree but an altogether different level of complexity.* A new type of geometry, even more alien to Euclid than that of Lobachevsky, developed in the first half of the 20th century. Mandelbrot coined the term 'fractal' (from the Latin *fractus* for broken [78]) for these bizarre and counter-intuitive geometrical objects, however as is usual in Mathematics some of the ideas had been floating around for a long time as we will see. It would be hard to give a precise definition as to what is a fractal and what isn't, however fractals share common features: they are often 'self-similar' meaning they contain copies of themselves within themselves; they are formed by infinite iterative processes; they have structure at all scales; and they have non-integer dimension. We will see what all of these terms mean via some archetypal examples, however I should emphasizes that the reason it is hard to define 'fractal' precisely (Mandelbrot himself tried to give a precise definition but it needed to be reworked and extended) is that it turned out that fractals are all around and can arise in many diverse contexts; alas I can only scratch

DOI: 10.1201/9781003455592-4

the surface, and a scratchy surface it is too, but hopefully you will see that an endearing property of fractals and one of the reasons they caught on outside the mathematical community is a democratization of mathematics - you can create your own fractals quite easily and draw pictures of mathematical objects no-one else has ever seen before.

We begin by going way back: Apollonius, of conic sections fame, was interested in the following problem (actually Apollonius considered a much more general problem [16], but we will focus on this special case): suppose there are three circles in the plane, all tangent to one another (in black on the right). Can you draw another circle that is tangent to all three?

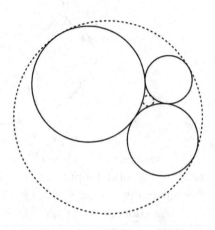

Apollonius knew that there are in fact two such circles, shown dashed, and he was interested in *constructing* these tangent circles, however Descartes had a more modern approach: he derived the following elegant formula, where k_1, k_2, k_3 are the curvatures of the three given circles and k_4 is the curvature of the new circle we seek:

$$(k_1 + k_2 + k_3 + k_4)^2 = 2(k_1^2 + k_2^2 + k_3^2 + k_4^2), \qquad (4.1)$$

which seems self-contradictory when you say it out loud: 'the sum of the curvatures squared is equal to twice the sum of the curvatures squared' (there is a vital pause needed). This is known as Descartes' 'kissing formula', and has the novel sign convention that if a circle is external to the other three then we write its curvature as negative; in the example above the three given circles have curvature $5, 8, 12$, and subbing into (4.1) and solving for k_4, we see the two new circles have curvature -3 (the large outer one) and 53 (the small inner one, remember small circles have large curvature). Since Descartes' kissing formula is quadratic, this explains why there are *two* new circles tangent to the three given circles.

Apollonius's construction of the two tangent circles is unfortunately lost to us, all we have are the writings of Pappus from several centuries later, but building on it Pappus had the following construction: take two circles, one inside the other and touching; then in the arc of space between them, put lots of circles all touching one another and the two original circles:

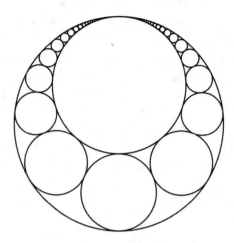

This is called a 'Pappus chain', and there is a subtle point to make here: Pappus said to put lots of circles in the space, lots and lots of circles, as many circles as you like, but he did not say 'suppose there are *infinitely many* circles in the space'. Pappus and other mathematicians from this era were very wary about infinity, and we will dwell on this some more in Chapter 9.

Actually there is a nice generalization, due to Steiner in the 1850s and known as a 'Steiner chain': rather than the two original circles touching, if we allow them to be slightly offset from one another then there is room to go 'round the back' and fill the whole region with circles, for example on the right. It turns out that the centers of those smaller circles filling the space all lie on a curve (shown dashed), and that curve is precisely an ellipse with foci at the

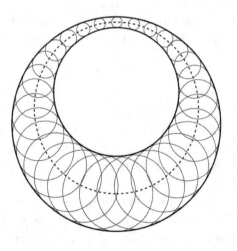

center of the two original circles; how elegant! Apollonius would have loved that. What's more, if the original two circles are outside one another, than the family of tangential circles have centers that lie on a hyperbola.

Leibniz, writing to a colleague in 1706 [101], went a step further: in the Pappus chain above there are spaces between the tangential circles,

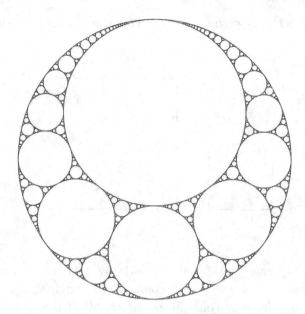

Figure 4.1: The Apollonian gasket, an early example of a fractal [135].

and in each space you can put a circle which is tangential to three circles, but then that creates new spaces, and so in those spaces you can put new circles which then creates new spaces and so on. Unlike Pappus, Leibniz had no hesitation in saying *imagine proceeding to infinity in this way*, thus describing what could be the first example of a fractal, called the 'Apollonian gasket', see Figure 4.1 (although he would not have described it as such of course; actually Leibniz's construction was a little more complicated than what I have described here).

What makes this a fractal, as opposed to the images drawn previously? The Apollonian gasket has 'structure on all scales', by which I mean if we zoom in on the space between circles then we see more circles, and if we zoom in on the space between those circles, we see yet more circles, and so on without end; *great fleas have little fleas upon their backs to bite 'em, and little fleas have lesser fleas, and so ad infinitum* (as de Morgan said). But you could argue that if I zoom in on a portion of a line, I'll see a portion of a line; is a line a fractal? A more formal test is needed for when something is a fractal, and for that we use the notion of dimension.

Let's start with a square, and to be clear we mean *including* the area inside the square. If you scale the length of a side by a factor of

2, then you get a new square which contains 4 copies of the original square:

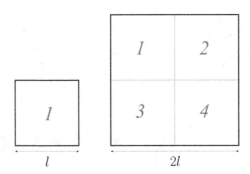

If you were to scale the length of a side by a factor of 3, you would get a new square that contains 9 copies of the original, and in general if you scale a side by n you will get n^2 copies of the original. This is of course because the square is an area and areas are 2-dimensional. If you had a cube, and scaled a side by a factor of 2, you would get a new cube that contains 8 copies of the original cube; more generally if you scaled a side by a factor of n you would get n^3 copies of the original, and this is because the cube is a volume and volumes are 3-dimensional. This might sound like I am stating the obvious, but this way of thinking about dimension, due to Hausdorff (1918), will enable us to think about the dimension of a fractal [40]. We could write it more carefully like this: the dimension of an object is d if

$$\left(\begin{array}{c} \text{number of copies of original} \\ \text{in scaled version} \end{array} \right) = \left(\begin{array}{c} \text{amount by which} \\ \text{a side is scaled} \end{array} \right)^d . \quad (4.2)$$

In 1915 Sierpiński published a paper describing an infinite iterative process on a triangle and in the following year a similar process on a square (actually his method in the first paper was to describe an increasingly wrinkled curve, and in the second he didn't include a picture; both papers are quite short, coming in at around 700 publications Sierpiński is not far off the output of Euler). We will describe the process on the square.

Let's start with a filled black square, and think of it as made up of a 3×3 grid of smaller squares. Step 1 is: remove the middle square.

You can see there are now 8 black squares around the empty square in the middle. Step 2 is: remove the middle square from each of these.

Now in each of those 8 squares there is a middle white square surrounded by 8 smaller black squares. Step 3 is: remove the middle square from each of them.

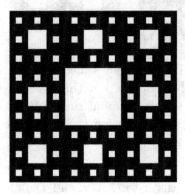

Just like Leibniz and his circles, the instruction is to keep repeating the same steps again and again, removing the middle square to create smaller squares and then removing their middle squares. Here's what it looks like after two more iterations:

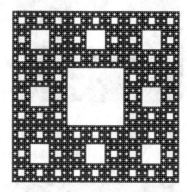

The Sierpiński square (also known as the Sierpiński carpet) is what happens after you perform this same step *infinitely many times.* What is the dimension of this object? Well look at the bottom left corner, which I have removed momentarily:

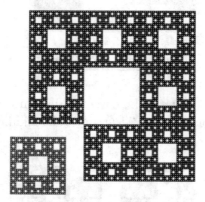

If you took this small image and scaled the side up by a factor of 3, you would get the whole image, but this *only contains 8 copies* of the smaller image; the 8 squares around the central white square. As such the equation defining the dimension (4.2) would read

$$8 = 3^d.$$

Clearly d cannot be 1, that's too small, and d cannot be 2, that's too big. Instead the dimension of the Sierpiński square is about 1.8928 (or more precisely $\log 8/\log 3$). So the Sierpiński square is more than a line, but less than an area; it exists somewhere in between these two everyday intuitive notions. This now is the main defining property of a fractal: non-integer dimension. We also see what is known as 'self-similarity', where the complete object contains copies of itself within it.

It gets stranger: you notice I needed less ink to draw the fifth iteration compared to the previous; how much ink would I need to draw the final version, with infinitely many iterations? None at all! Mandelbrot [79] described these constructions as 'gaskets', because they are like the gasket in a car engine: a sheet of material separating the two halves of the engine block which has several holes of various sizes punched into it. Let's continue this analogy and suppose you started with a square of material and punched a square hole in the middle like in the first iteration above. Then you kept on punching square holes of smaller and smaller size just like we described. When you are finished, the sheet of material will weigh nothing at all; the area of the Sierpiński square is zero!

The wonderful thing about fractals is you can make your own - let's take a 5×5 grid of black squares, and color some of them white. I have chosen a capital 'B', because it reminds me of my favorite fractal joke: what does the 'B' in Benoit B. Mandelbrot stand for? Benoit B. Mandelbrot.

Now for each of the little black squares we divide it into a 5×5 grid and color some of them white in the same pattern; then in each of those black squares divide and color white in the same pattern and so on forever.

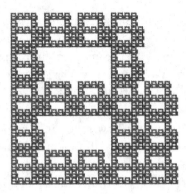

This is again a fractal, and its dimension is $\log 19/\log 5$ which is about 1.8295 (if you take a small square and scale a side by a factor of 5 you will get 19 copies of the same square). In general, if you divide a square into an $n \times n$ grid, and color m of the squares white to form a pattern, then the resulting fractal will have dimension $\log(n^2 - m)/\log n$ [19].

What we have described in the previous section could be thought of as a 'geometrical' way of generating fractals, however the most famous fractal of all, the Mandelbrot set, is generated in a more 'numerical' way, by repeatedly applying a rule to a number and seeing what happens. As this process is called 'iteration', the sequence of numbers you get by applying the rule are called 'iterates'. You will not be surprised to hear that this idea of iteration on numbers has been around for a very long time in many different cultures, from the Pythagoreans in Greece to Brahmagupta in India [109]. The typical notation goes like this: let the first number in the sequence be x_0, the next be x_1 and so on. Letting x_n be the nth number, the next one x_{n+1} is derived from the previous ones according to some rule. The main question to ask then is: what happens to the sequence of numbers as we repeatedly apply the rule?

A well known example is the Fibonacci sequence: starting with $x_0 = 1$ and $x_1 = 1$, let

$$x_{n+1} = x_n + x_{n-1},$$

in other words, the next number in the sequence is the sum of the previous two. So to find x_2 the rule says $x_2 = x_1 + x_0$ which is equal to 2, and then $x_3 = x_2 + x_1$ which is equal to 3; then $x_4 = 5$, $x_5 = 8$ and so on. Among the many interesting properties of the Fibonacci sequence, the ratio of successive terms in the sequence approaches the so-called 'golden ratio', denoted ϕ, to which some people have attached almost mystical significance [89].

The numbers in the sequence don't have to be integers, for example the 'logistic map' is the rule

$$x_{n+1} = \lambda x_n (1 - x_n)$$

and the question is: what happens to the sequence for different values of λ? For the Mandelbrot set, we need to extend the type of number we can iterate yet further, to include complex numbers. In Chapter 6 I will introduce complex numbers in the traditional manner, however for the sake of this chapter there is another way to view complex numbers which suits us much better: *as points in the plane that can be multiplied*.

We will follow the convention to denote a complex number with z, and since we are thinking of complex numbers as points in the plane, we can write $z = (a, b)$ where a, b are the coordinates of the point. We can clearly add complex numbers to get another complex number (if $z_1 = (a, b)$ and $z_2 = (c, d)$ then $z_1 + z_2 = (a + c, b + d)$), and we can scale by a factor to get another complex number (if $z = (a, b)$ and k is a scale factor then $kz = (ka, kb)$), but now we can also define a meaningful way of multiplying complex numbers to get another complex number: if $z_1 = (a, b)$ and $z_2 = (c, d)$ then

$$z_1 \times z_2 = (ac - bd, ad + bc). \tag{4.3}$$

Actually all we really need is to be able to square complex numbers: if $z = (a, b)$ then

$$z^2 = (a^2 - b^2, 2ab). \tag{4.4}$$

Some examples are below, but this is all we need to know about complex numbers in order to define the Mandelbrot set.

In 1977 Benoit B. Mandelbrot published *Fractals: form, chance and dimension* [78]. It was something of a call to arms: Mandelbrot put forward his thesis that the intricate forms of nature cannot be adequately captured by the circles and cones of conventional Mathematics, but instead can be described by a class of objects for which he coined the term 'fractal'. He considered a variety of phenomena from the natural world, such as the lengths of coastlines and rivers, the distribution of celestial objects like stars and galaxies, turbulence and curdling, lunar craters and soap bubbles, all modeled by objects with non-integer dimension (the Sierpiński square is Plate 167). What is not in the book is the fractal named after him, that came in his 1982 *The fractal geometry of nature* [79]. In it, Mandelbrot refers to the work of Fatou and Julia, two French mathematicians who studied iterations on complex numbers in the early 1900s. For example, taking the logistic map mentioned above but now letting the numbers being iterated be complex:

$$z_{n+1} = \lambda z_n (1 - z_n),$$

we can make a transformation to give the rule

$$z_{n+1} = z_n^2 - \mu. \tag{4.5}$$

Mandelbrot referred to this as 'the μ-map', and he defined the following set: for different values of μ, start with $z_0 = (0,0)$ and keep iterating according to the μ-map to get a sequence of complex numbers; keep going until you can decide if the sequence is staying bounded (within some circle around the origin) or becoming unbounded (will leave any circle around the origin eventually). Those μ which lead to a sequence staying bounded are *in* the set, those which lead to the sequence becoming unbounded are *not* in the set (actually the original definition of the set is more complicated and hard to compute, however this more amenable description is completely equivalent [40]). Let's do some examples.

Suppose μ is the complex number given by the point $(1,0)$. If z_0 is $(0,0)$ then
$$z_1 = z_0^2 - \mu = (0,0) - (1,0) = (-1,0),$$

where we are using our rule in (4.4) for squaring complex numbers. We now use z_1 to find z_2:

$$z_2 = z_1^2 - \mu = (1,0) - (1,0) = (0,0).$$

We have got back to z_0 again, and repeatedly iterating will just go backwards and forwards between $(-1,0)$ and $(0,0)$. This sequence is bounded, and therefore $(1,0)$ is in the set.

Now let us choose $\mu = (-0.3, -0.3)$. Again $z_0 = (0,0)$ so

$$z_1 = z_0^2 - \mu = (0.3, 0.3), \quad z_2 = z_1^2 - \mu = (0.3, 0.48), \quad z_3 = \dots$$

but this quickly becomes tedious. Instead we can draw a picture: a dot at each point in the sequence, joined by a line to bring out the order, on the left below.

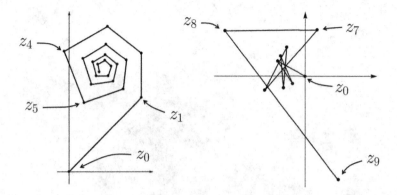

We can see the sequence of points spiralling inwards and so this sequence is bounded; the point $(-0.3, -0.3)$ is in the set. On the other hand, if we started with $\mu = (0.75, -0.4)$, then the sequence of points spirals outwards and becomes unbounded (on the right above); the point $(0.75, -0.4)$ is not in the set. A natural question to ask is: what does the boundary between the points in the set and the points not in the set look like?

If we divide the complex plane into a grid of very many points, and work our way through them one at a time deciding if that point is in the set (color it black) or not (color it white) we get the following picture:

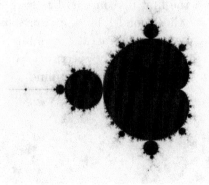

(This is actually the reflection of Mandelbrot's original drawings, Plate 188 in [79], as the convention today is to define the μ-map as $z_{n+1} = z_n^2 + \mu$). The black region is the Mandelbrot set, and it looks a bit underwhelming at first, a vaguely heart-shaped blob with some blobs attached; however as you look closer and closer at the boundary of the set, *extraordinary fractal riches* are revealed, as Mandelbrot put it. Here is just one example, where I have zoomed in on a part of the boundary:

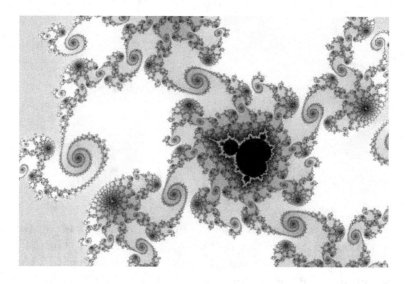

Some nuance is added to the pictures: rather than just coloring a point white if its iterates become unbounded, we can add a grayscale depending on how quickly it becomes unbounded. There are lots of places online where you can draw your own pictures of the Mandelbrot set, these came from [27], and here's the thing: if you choose a random part of the boundary of the Mandelbrot set and zoom in numerous times, there is a good chance that you are the first human being to ever see that part of the set; perhaps I am the first person who has ever seen the images on this page (and you are the second)! These pictures are really not doing it justice: there is a plethora of fractal art online, some of it hugely sophisticated and creative. I strongly encourage you to open YouTube and search 'fractal zoom', it will blow your mind.

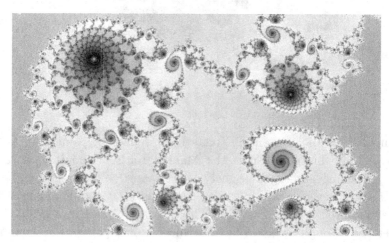

Mandelbrot used very naturalistic terms to describe his fractals, 'corona', 'sprout' and 'root', and the online community followed suit; parts of the Mandelbrot set are called 'seahorse valley', 'elephant valley' and 'the needle' for example [13]. As you zoom in you come across copies of the Mandelbrot set within itself, like in the image above, called 'minibrots', but it is harder to calculate the dimension using the scaling trick we did earlier. Nonetheless it has been shown that the Mandelbrot set, being a region of the plane, has dimension 2 (as we should expect), but the *boundary* to the Mandelbrot set *also* has dimension 2; both the region and its boundary have the same dimension, bizarre!

Mandelbrot's big idea seemed to catch the zeitgeist, and fractals are one of the rare instances when a mathematical concept has crossed over into popular culture. Part of the reason for this is that Mandelbrot was right and many natural phenomena can indeed be described by fractals, such as trees, lungs, lightening, blood vessels and so on, and this is something people can easily relate to. Another reason is that the images of fractals are cool and bizarre, earning a place in any discussion of the relationship between mathematics and art [45, 89] but also the accessibility of fractal art: in the 80s and 90s the home computer was becoming the norm, and anyone could write some simple code to generate their own fractal images. Finally the development of fractal geometry was closely linked to another important development in 20th-century mathematics: chaos, which we will see again in Chapter 10. Together fractals and chaos gave us a whole new lexicon of terms such as the butterfly effect, basins of attraction and strange attractors. If we remember that the word 'geometry' literally means 'measuring the Earth', it is perhaps ironic that this fractal geometry, which is so far removed from that of Euclid, was developed to provide a better description of the natural world.

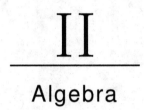

Algebra

CHAPTER 5

The Beginnings

I F YOU think of literally any image in the media that is supposed to represent mathematics, it will most likely involve equations; either floating in the ether or being written on a blackboard or something. Algebra, in the sense of symbols and equations, is synonymous with Mathematics in most people's minds, but what we would today recognize as Algebra is a relatively recent development (16th century as we will see). Instead Algebra has a long pre-history which I cannot describe in detail (so much to say! Alas, Statement A) but will sketch in this chapter, and when I say pre-history, I mean it – we begin way way back.

Discovered in the 1950s in modern day Democratic Republic of the Congo, buried under a pre-historic volcano, the *Ishango bone* (above [136]) is the bone of an unknown animal with groups of notches arranged in three columns along it, 168 marks in all [41]. Dating methods can be disputed but the current accepted age for this artefact is 20,000 years old, making this one of the oldest known mathematical objects. It is reasonable to assume that simple tally systems like this were prevalent in very many places at many different times, but unfortunately physical records have been largely lost. However, carved in stone and inked onto papyrus in ancient Egypt at least 5000 years ago, early *hieroglyphic* writing [16] used a tally-like number system: the symbol for 1 was a vertical dash |, 10 was a little hoop ∩, 100 a curly 9 𝟿 and so on, and these were simply repeated as many times as needed; for example the number 325 was written 𝟿𝟿𝟿∩''¦¦ (here and in what follows I will order the symbols

Figure 5.1: The *Plimpton 322* cuneiform tablet c. 1800 BCE, which consists of tables of numbers related to Pythagorean triples [61].

in the way we would understand them easiest). About 4000 years ago the Egyptians started using 'ciphers', where each digit was represented by its own symbol, known as the *hieratic* system; in this system 325 is now represented as ⟋ʼⴷ᠕.

Also about 4000 years ago the Babylonians were using a wedge to make a mark in clay, and since the Latin word for 'wedge' is *cuneus* this system of writing is called *cuneiform*; see Figure 5.1. A thin vertical wedge Ⴤ was 1, a wider sideways wedge ⊀ was 10, and the number 24 would be represented as ⟪ᛦ. The Babylonians took a big step toward a 'positional' system which we would be familiar with (for example, when we write '22' we are using the same symbol twice, but we know the first '2' is for tens, our base, and the second '2' is for units); they used base 60, so when they wrote Ⴤᛦ Ⴤ this meant 2 lots of 60 and 1 lot of 1, so this was a short way of writing 121. They also introduced, just over 2000 years ago, a 'placeholder', which was almost like a zero; they used the mark ⤴ to denote an empty slot, however it was not complete: ᛦᛦ⤴ᛦᛦ could represent $3(60)^2 + 0(60) + 2(1)$ or $3(60)^3 + 0(60)^2 + 2(60)$ for example.

In the time of Euclid and Archimedes the Greek *Ionian* number system was based on their alphabet, with (capital) letters representing numbers 1 to 9 (A,B,Γ being 1,2,3 and so on), multiples of 10 (I,K,Λ being 10,20,30) and multiples of 100 (R,Σ,T being 100,200,300). Running out of letters, the same symbol was re-used for higher powers of 10 but with a dash to distinguish them, so for example 3,000 was ′Γ and 3,223 was

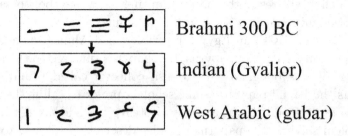

Figure 5.2: The evolution of the first five numbers of the Hindu-Arabic system that we use today [68].

ΓΣΚΓ, a rudimentary positional system. At about the same time Chinese writing used *rod* numerals which were positional, but there was also a multiplicative system: there were symbols for 10 | and 100 ☻, and then attached to them were the numbers multiplying them; so if)(was 8 and ⇪ was 6 then 868 was ☻ᵢ)ᶜ.

Finally, by the 8th century CE a fully positional decimal based system arose in the Hindu world of India and South East Asia, using just nine symbols and a symbol for zero. This was then adopted by the Islamic mathematicians in the medieval period (see Figure 5.2 and al-Khwārizmī below) before gradually making its way to Europe and evolving into the number system we use today, the 'Arabic' system, which some scholars more carefully call 'Hindu-Arabic'. One of the key figures to bring this number system to Europe was Leonardo of Pisa, also known as Fibonacci (of Fibonacci number fame), whose *Liber Abaci* of 1202 described what he had learnt on his travels in the East as a merchant. It is said he would go to the market and win bets with the local traders by challenging them to see who could do a computation faster, his use of Hindu-Arabic numerals giving him the winning edge. It was slow to catch on; in the 15th-century traders are still using Roman numerals or even a hybrid system like $III^m IV^c XXI$ for 3,421 [114].

Based on these number systems our ancestors developed the elements of arithmetic, but also began answering more complex problems which we would today consider Algebra; an important point to note however is that (with a few exceptions) these problems and their solutions were stated entirely in words and numbers, without symbols for the unknowns, and as such I will resist the urge, tempting as it may be, to say 'this problem is equivalent to $x^2 + 4x = \ldots$' or similar.

Around 1550 BCE, an Egyptian scribe called Ahmes wrote a series of problems on a length of papyrus; this papyrus survived to the modern age where it was bought by a Scottish collector called Rhind so it is known as the Rhind papyrus (some scholars prefer to call it the Ahmes papyrus after the person who actually wrote it). Intended as instructional for problem solving, the papyrus includes descriptions of how to divide bread and beer equally between people, but also several problems that don't make reference to a specific substance, instead using the term *aha* meaning 'heap'. For example Problem 30 [16] asks: if heap plus two thirds heap plus one half heap plus one seventh heap is thirty seven, what is heap? We would call these 'linear' problems today. After finding the solution, Ahmes then goes to the effort of showing his solution is correct; if the papyrus was intended as a set of exercises for trainee scribes it is nice to see Ahmes giving the same advice we still give students today: check you answer! There are also geometrical problems, for example Problem 51 calculates the area of an isosceles triangle, but no formula or symbols are used; there is a *specific* triangle in mind. The Moscow papyrus, slightly older than Ahmes, asks for the volume of a truncated square pyramid if the edges of the upper and lower bases are two and four units respectively and the height is six units; the method for constructing the solution in this specific instance is given but again no general formula or symbols.

The Babylonians went further, solving problems like [16] what is the side of a square if the area less a side is $14\frac{1}{2}$? If the sum and product of two numbers is given, what are those numbers? We would call these 'quadratic' problems today, but they even solved large classes of 'cubic' problems. They divided these problems up into different classes because they were careful to only consider problems with positive solutions and positive coefficients, largely because these problems were viewed geometrically (we give an example below) and so negative lengths and areas made no sense; having said that they were happy to add an area and a length for example, whereas others were not.

In ancient China there was a text called *Jiuzhang suanshu* (Nine Chapters on the Mathematical Art) which was developed and added to over many years, perhaps from as early as 1000 BCE to 200 BCE, and Chapter 8 begins with the following problem [68]: *there are three classes of grain, of which three bundles of the first class, two of the second and one of the third make 39 measures. Two of the first, three of the second and one of the third make 34 measures. And one of the first, two of the second, and three of the third make 26 measures. How many measures of grain are contained in one bundle of each class?* The text instructs us to write the numbers 3, 2 and 1 for the bundles and 39 for the measure in a column on the right, and the other numbers in the middle and left, to give the table of numbers on the left below:

$$
\begin{array}{ccc}
1 & 2 & 3 \\
2 & 3 & 2 \\
3 & 1 & 1 \\
26 & 34 & 39
\end{array}
\qquad
\begin{array}{ccc}
1 & 0 & 3 \\
2 & 5 & 2 \\
3 & 1 & 1 \\
26 & 24 & 39
\end{array}
\qquad
\begin{array}{ccc}
0 & 0 & 3 \\
0 & 5 & 2 \\
36 & 1 & 1 \\
99 & 24 & 39
\end{array}
\qquad (5.1)
$$

Then we are told to multiply the middle column by 3, and subtract from each entry 2 times the right column, to get the table in the middle; note we have introduced a 0. Performing several operations like this gives the table of numbers on the right, and now the solution can be found by starting with the leftmost column and working rightwards. It was a good 2000 years before this method was rediscovered in Europe by Newton and Gauss, although it was in use by Chinese mathematicians since the time of Euclid. One of the problems in Chapter 8 even considers 5 equations with 6 unknowns (I mean Chapter 8 of *Jiuzhang suanshu*; Chapter 8 of this book only has 4 equations in 6 unknowns (8.3)). That problem has infinitely many solutions, but only one is given. This is quite typical of pre-modern Algebra: where there is more than one solution only one is given, the one with all positive or whole numbers.

A slight exception is in Indian mathematics, where working with positive and negative numbers, and more than one solution to a problem, was more common, as well as a more general approach: Brahmagupta (7th century CE) describes what to do when a 'square' and a 'simple unknown' are equal to an 'absolute number'; he says [68] *take absolute number on the side opposite to that on which the square and simple unknown are. To the absolute number multiplied by four times the coefficient of the square, add the square of the coefficient of the unknown; the square root of the same, less the coefficient of the unknown, being divided by twice the coefficient of the square is the value of the unknown* (this is really pushing my 'only use symbols when they were used by the author' mantra). Bhāskara (12th century) recognized that a quadratic problem could have *a two-fold value for the unknown quantity*, but, sensing things start getting complicated when you follow this chain of thought, says nothing more than *this holds in some cases*.

We get a sense of many different cultures attacking broadly similar problems we would today describe as linear, quadratic and cubic, and possibly even with several unknowns. Also you can see how difficult it is to make progress without symbolic expression, for example that last quote by Brahmagupta would be easily recognized if written in modern notation (see (7.1)). The next figure we discuss was possibly the first to develop a symbolic representation of unknowns: Diophantus.

Almost nothing is known about Diophantus himself (Alexandria, probably mid 3rd century CE) but his gift to the ages is his *Arithmetica* (the word 'arithmetic' meant more 'number theory' rather than 'computation' at the time): 13 books, not all of which survived, consisting of a long list of problems and step by step instructions on how to solve them, the problems themselves phrased in the abstract and making no reference to bread or grain. In it, he introduced the following terms and symbolism: a square (which is when a number is multiplied by itself) is denoted Δ^Y, the side of the square being ς; cubes are K^Y, squares of squares are $\Delta^Y\Delta$ and so on; numbers are M^o. The multiples of these terms are written using the Ionian system described previously, so for example 'three times a square plus twice the side plus twelve' would be written $\Gamma\Delta^Y B\varsigma KBM^o$. The symbol ⋔ meant everything afterwards had a minus in front, and there was even the use of χ to denote reciprocals, so for example the reciprocal of a square was $\Delta^Y{}_\chi$. Now we could write 'a cube plus thirty minus three times the side' as $AK^Y\Lambda M^o$ ⋔ $\Gamma\varsigma$. You might enjoy trying to figure out what $IB\Delta^Y\Delta\Gamma\varsigma$ ⋔ $\Gamma K^Y AM^o$ corresponds to; it might seem cumbersome at first, but only because we are used to another system. While this symbolic system was not picked up by other mathematicians (as far as we know), his influence in phrasing problems in a more abstract and number theoretical way had a long reach....

If you were to write something, would you hope that in a thousand years it would inspire one of the leading mathematicians of the day to conjecture one of the most famous open problems in Mathematics? Hypatia, who we met in Chapter 1, wrote a commentary on Diophantus's *Arithmetica* around 400 CE [68] and that commentary was translated into Arabic in perhaps the 10th century; this Arabic version then found its way into Europe where it was translated into Greek and Latin in 1621, finally ending up in the hands of Pierre de Fermat shortly afterwards. Book II Problem 8 of the *Arithmetica* asks to divide a square into

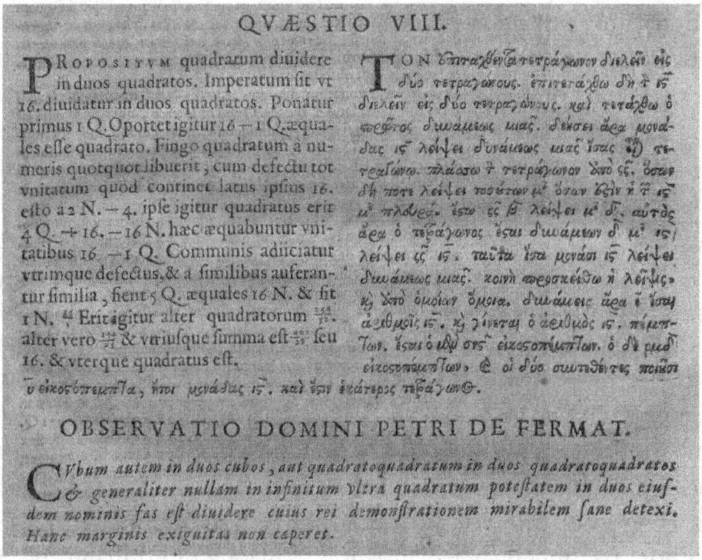

QVÆSTIO VIII.

PROPOSITVM quadratum diuidere in duos quadratos. Imperatum sit vt 16. diuidatur in duos quadratos. Ponatur primus 1 Q. Oportet igitur 16 − 1 Q. æquales esse quadrato. Fingo quadratum a numeris quotquot libuerit, cum defectu tot vnitatum quod continet latus ipsius 16. esto a 2 N. − 4. ipse igitur quadratus erit 4 Q. + 16. − 16 N. hæc æquabuntur vnitatibus 16 − 1 Q. Communis adiiciatur vtrimque defectus, & a similibus auferantur similia, fient 5 Q. æquales 16 N. & fit 1 N. 4 Erit igitur alter quadratorum 16/25. alter vero 144/25 & vtriusque summa est 400/25 seu 16. & vterque quadratus est.

ΤΟΝ ἐπιταχθέντα τετράγωνον διελεῖν εἰς δύο τετραγώνους. ἐπιτετάχθω δὴ ὁ 16 δηλεῖν εἰς δύο τετραγώνους. καὶ τετάχθω ὁ πρῶτος δυνάμεως μιᾶς. δήσει ἄρα μονάδας 16 λείψει δυνάμεως μιᾶς ἴσας τῷ τετραγώνῳ. πλάσσω τὸν τετράγωνον ἀπὸ 16, ὅσων δή ποτε λείψει ποσῶν μ' ὅσον ἐστὶν ἡ Γ 16 μ' πλήθει. ἔστω ἐς β λείψει μ' δ'. αὐτὸς ἄρα ὁ τετράγωνος ἔσται δυνάμεων δ' μ' 16 λείψει ἐς 16. ταῦτα ἴσα μονάσι 16 λείψει δυνάμεως μιᾶς. κοινὴ προσκείσθω ἡ λείψις, καὶ ἀπὸ ὁμοίων ὅμοια. δυνάμεις ἄρα ε ἴσαι ἀριθμοῖς 16. καὶ γίνεται ὁ ἀριθμὸς 16. πεμπτων. ἔσται ὁ μὲν ὅτε εἰκοσιπεμπτῶν, ὁ δὲ μ''δ εἰκοσιπεμπτῶν, καὶ οἱ δύο συντεθέντες ποιοῦσι

υ εἰκοσιπεμπτῶν, ἤτοι μονάδας 16, καὶ ἔστιν ἑκάτερος τετράγων@.

OBSERVATIO DOMINI PETRI DE FERMAT.

CVbum autem in duos cubos, aut quadratoquadratum in duos quadratoquadratos & generaliter nullam in infinitum vltra quadratum potestatem in duos eiusdem nominis fas est diuidere cuius rei demonstrationem mirabilem sane detexi. Hanc marginis exiguitas non caperet.

Figure 5.3: Book II Problem 8 of the *Arithmetica*: to divide a square into two squares, from the 1670 version [28], with Latin and Greek. Note the 'observatio' just after the problem; this is the marginal note of Pierre de Fermat, with the famous line *hanc marginis exiguitas non caperet*: which this margin is too narrow to contain.

two squares, which probably lead to Fermat asking if a cube could be divided into two cubes, and a quartic into two quartics. Fermat made the following conjecture, which he wrote into the margin of his copy of *Arithmetica*: that it is impossible to write a cube as the sum of two cubes, or a fourth power as the sum of two fourth powers, and so on. Then quite famously he wrote: *I have discovered a truly marvellous proof of this, which this margin is too narrow to contain.* After his death, his son published a new version of Diophantus's *Arithmetica* [28], and included all his father's marginal notes in the text (see Figure 5.3), including the one just mentioned. This conjecture became known as 'Fermat's Last Theorem', and resisted the concerted efforts of mathematicians (including Sophie Germain, who we will meet in Chapter 12) for several centuries before finally being proved by Andrew Wiles in the 1990s; I thoroughly recommend Simon Singh's description [104].

Lagrange said *Diophantus can be considered the inventor of Algebra* [53], but not everyone would agree: in the *Arithmetica* we see the first formal use of symbols, and the equivalent of powers higher than three for the first time, but no exponential notation or symbols for operations; the problems are abstract but still specific, rather than a general treatment of problems of a certain type; he often treats problems with more than one solution, but only gives one; finally the order of presentation seems bizarre and disconnected and it is not a systematic exposition on algebraic methods [16]; unlike al-Khwārizmī.

One of the first scholars to be called to the House of Wisdom in Baghdad, Muḥammad ibn Mūsā al-Khwārizmī, who died perhaps in the early 9th century, wrote two very important books in the development of Mathematics. The first is *Kitāb al-jam'wal tafrīq bi ḥisāb al-Hind*, the Book on Addition and Subtraction after the Method of the Indians. In it he described the nine numerals of the Hindu-Arabic system and a circle for zero, then went on to show how to write any number in this system but also how to quickly perform the operations of addition, multiplication, extracting roots and so on. In fact he described this so completely that when the text entered Europe people thought he had invented these methods, to the extent that performing these operations became known as *algorismi* in a corruption of al-Khwārizmī (the Latin translation begins *Dixit Algorismi...*, Al-Khwārizmī says...), which developed over the years into the modern word *algorithm* meaning any general step-by-step process.

Even more significant was his *Al-kitāb al-muḫtaṣar fī ḥisāb al-jabr wa-l-muqābala*, the Condensed Book on Calculation by Completing and Balancing; for short it was known as *al-jabr* and it is from this we get our word Algebra. Al-Khwārizmī considered three kinds of numbers: squares, roots (of squares, i.e the unknown), and absolute numbers. He described six different classes of quadratic equations: squares equal to roots, roots and numbers equal to squares, and so on; this was because he only countenanced positive numbers, and these are the six types of quadratics which then guarantee positive roots. Each example and its solution was described in words, but then was proved using geometrical arguments. A good example [68] would be a problem of the type 'squares and roots equal to numbers', which he stated as *what must be the square which, when increased by ten of its own roots, amounts to thirty-nine?* His geometrical proof that the word solution he provided is correct works like this:

We draw a square, whose side is the number we seek (I am resisting the urge to label it x or something, but you go ahead!). We then extend that by two rectangles of side five (i.e. half of ten).

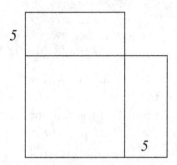

The total area of these regions must be thirty-nine from the statement of the problem, but note there is a little notch in the top corner. If we 'complete the square', by adding the missing square which must have area twenty-five, then this large square will have area sixty-four, which means it has side eight, and finally the length we seek must be three.

In some respects this is not much of an advance on the Babylonians who used similar 'complete the square' techniques, and there are no symbols used (Diophantus had not yet been translated into Arabic); everything was in words, even the numbers as I have written above, and no general treatment but specific examples. Nonetheless, the analysis in *al-jabr* is more systematic, dividing quadratics into an exhaustive list of different classes and treating each in turn; detailed explanations are given showing how to solve each class of problem and then formal geometrical proofs are provided to convince the reader that the method works. It is for these reasons that al-Khwārizmī is considered, along with Diophantus, as one of the founding figures in Algebra, before Renaissance mathematicians took the Algebra ball and ran with it.

Being a mathematician in 16th century Italy was precarious work. While you might be employed at a university (the University of Bologna in 1500 had 16 astronomers [114]) you needed to stay in the favour of your employer and you could be challenged at any time to a competition for your position; a rival would present a list of problems for you to solve, and you would do the same to them, and whoever could solve the problems in the most impressive manner would win the position, a mathematical duel

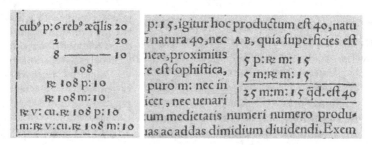

Figure 5.4: Two excerpts from Cardano's *Ars Magna*, 1545 [44]; on the left we see the solution to the equation *cubus più 6 rebus aequalis 20* (the cube of the unknown plus 6 of the unknown equals 20), given as *the cubed root of the square root of 108 plus 10 minus the cubed root of the square root of 108 minus 10* (also known as '2').

if you like. For this reason, if you came up with a new method to solve a tricky problem then it was in your interest to keep it to yourself, as an ace card up your sleeve to pull out if you were challenged (I will never again complain about the 'publish or perish' culture that exists today). Linear and quadratic problems were well understood (thanks to *al-jabr*) but the cubic equation was still a mystery (the work of Babylonians and Islamic scholars such as Omar al-Khayyāmī and Sharaf al-Dīn al-Tūsī [68] being unknown at the time). The story goes [109] that Scipione del Ferro, perhaps around 1515, found the solution to the cubic of the class 'cubes plus roots equals numbers', and in 1535 Niccolò Tartaglia, while preparing for a contest with one of del Ferro's colleagues, also solved cubics of this class plus perhaps those of the class 'cubes plus squares equals numbers'. Hearing of this, Girolamo Cardano begged Tartaglia to share his secret for a book Cardano was writing; eventually Tartaglia agreed, but only under the strict condition that Cardano would not publish it until he (Tartaglia) published first. In fact because the solution method is hard to remember, Tartaglia shared it in the form of a poem ("bad Italian verse" as Lagrange said [53], and he would know: he was born Giuseppe de Lagrangia). Mnemonics in the form of songs or poems are common, I recommend the quadratic formula sung to the tune of 'pop goes the weasel'.

Cardano grew impatient, and when he found out that del Ferro had in fact priority over Tartaglia [68] he went ahead and published his *Ars Magna*, The Great Art, in 1545, see Figure 5.4 (notation at this time was essentially abbreviations, 'p' and 'm' from *più* and *meno*, plus and minus, and roots with an R from *radice*, radix). In it, he gave the solution of all classes of cubics that have positive coefficients and admit positive solutions, and in fairness he gave credit where it was due to Tartaglia (he also opens the book by giving credit to al-Khwārizmī and Fibonacci), but this is a rare moment of humility; *Ars Magna* is full of braggadocio: Cardano describes himself as an *outstanding mathematician*, the book as *the perfect work*, and finishes with the statement *written in five years, may it last as many thousands*. Needless to say Tartaglia was not impressed, but the solution of the cubic is now sometimes known as 'Cardano's formula'. Personally I think this is a bit generous; Cardano describes several classes of cubic, and in each case describes a solution method in words, demonstrated by examples with specific numbers as coefficients and geometrical support, very much like *al-jabr*. As such there is still not a 'formula' which applies to *all* cubics, like the quadratic formula most readers will be familiar with. For that, we had to wait another generation for the Frenchman François Viète.

Figure 5.5: An excerpt from Viète's *Opera Mathematica*, 1646 [122], showing the general solution of the cubic.

Viète is a somewhat overlooked mathematician, but he was quite ahead of his time: he proposed that planets orbit the Sun in elliptical orbits (Chapter 11), he gave an infinite expression for π and summed infinite series [53], and he gave to Algebra what to my mind is its most immediately recognizable feature: letters of the alphabet to represent the unknowns, but also letters of the alphabet to represent the coefficients, so he could now finally give a formula for *all* quadratics or *all* cubics. In his 1591 *In artem analyticam isagoge*, Introduction to the art of analysis, he described an approach to problem solving which has become second nature to mathematicians since: you use symbols to represent the unknowns, and then try and write an algebraic equation to be solved for the unknowns. He used vowels for the unknowns and consonants for the coefficients, so for example A cubed $-3BA$ equals $2Z$ has solution (see Figure 5.5)

$$\sqrt[3]{Z + \sqrt{Z^2 - B^3}} + \sqrt[3]{Z - \sqrt{Z^2 - B^3}}, \tag{5.2}$$

and this is the first time we see the general solution for *all* problems of this type, rather than specific cases (like in Figure 5.4 where B is -2 and Z is 10). His notation is closer to ours in that he is using the $\sqrt{}$ sign (which was introduced by Christoph Rudolff in 1525, possible by an elaboration on the letter 'r' for 'radix') with $\sqrt{C}$ for cubed root, and the $+$ and $-$ signs [23], but note he is averse to adding numbers of different degree (how can we add a length to an area?) so he is at pains to make all the terms in a combination have the same dimension (the principle of 'homogeneity') which is why we see the words 'plano' and 'solido'. Nonetheless, Viète's inspired symbolism put Algebra on a new level of completeness, and was the kindling that lead, 30 years after his death, to a French Revolution.

There have been few marriages as happy and fruitful as that of Geometry and Algebra. An odd match, Geometry venerable and established, Algebra fresh and exciting, nonetheless this union gave birth to wondrous offspring. Presiding over the ceremony were two celebrants, Descartes and Fermat, like an awkward but well-meaning attempt at ecumenism. Descartes's *La géométrie* of 1637 (as described in Chapter 1), and Fermat's *Ad locos planos et solidos isagoge*, (Introduction to plane and solid loci, written in 1637 also but not published till 1679), both hit upon the same idea independently and at the same time: to use the new algebraic methods to study geometrical problems and vice versa, a method which came to be known as Analytic Geometry. Both Descartes and Fermat had read Apollonius and Viète deeply, and both made the connection between an equation containing two unknowns, the algebra part, to a curve in the plane, the geometry part. The difference between their approaches was their starting point: Fermat was inclined to begin with equations and relate them to curves, so he naturally started with linear and then quadratic equations, whereas Descartes began with curves and then tried to derive the equations that describe them, which were often of a higher degree. They both realized that if you have a pair of coordinate axes labeled x and y (like we described in Chapter 1), then any *linear* algebraic equation of the form

$$ax + by = c$$

was geometrically described as a line (and vice versa: any line could be captured by a linear algebraic relation between x and y), whereas any *quadratic* equation of the form [68]

$$ax^2 + 2bxy + cy^2 = 1 \tag{5.3}$$

is a conic section; indeed if $ac - b^2 > 0$ then this curve is an ellipse, whereas if $ac - b^2 < 0$ it is a hyperbola (see Figure 5.6). I encourage you to look back to Apollonius's definition of an ellipse from Chapter 1, and marvel at how much clearer and more compact is the algebraic description; now, for the first time in many centuries, Mathematics was stepping out of the shadow of the ancients and genuinely inventing again.

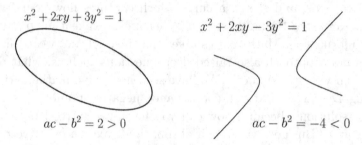

Figure 5.6: Classification of curves of the form (5.3).

You will also notice I am (finally!) writing equations, and reading *La géométrie*, the modern reader will see much that is familiar: Descartes was the first to use letters from the end of the alphabet, like x, y, z, for the unknowns and from the beginning of the alphabet for the parameters, and he uses exponential notation like a^3 and b^4 (but curiously stuck with aa for a^2, as did Gauss even in his 1827 *Disquisitiones*); the only thing that jars is he uses a funny backwards $\propto$ for the 'equals' sign, even though in 1557 Robert Recorde introduced the '=' sign based on a pair of parallel lines, for as he said, *Noe 2 thynges can be moare equalle* [23].

At the risk of hyperbole, this marriage of Geometry and Algebra really was a revolution; *this* was the key conceptual leap forward, propelling Mathematics into the modern era. It is perhaps hard for us to appreciate, since we have become so used to this mental framework, that in fact it pervades how we describe the world even in the day-to-day. Previously, an expression like $x^2 - 2x + 3$ was of interest only for its roots, but now

$$y = x^2 - 2x + 3$$

was so much more: as a curve in the plane the roots are when the curve crosses the x-axis, but we can also get a visual grasp on how the

y-coordinate is varying due to changes in the x-coordinate, and this was the beginning of the idea of a 'function', like a machine with an input (x) and an output (y), and the curve is the 'graph' of the function. While this idea is implicit in Descartes' and Fermat's writing it was Leibniz who first described functions and Euler who gave us the notation $y = f(x)$. But just switch on the news any day or open a paper and you will see this idea in action again and again: a graph of interest rates over time, or house prices or unemployment figures. Think of a friend or a sibling or a cat; can you imagine in your mind a pair of axes, with time on the horizontal axis and how well you got on with your friend/sibling/cat on the vertical axis, and sketch a curve which captures how your feelings changed over time? Humans are visual creatures, we like using pictures to convey information; Mathematics takes that instinct and weaponizes it.

This new approach also raised deep questions such as: what do we mean when we say 'number'? We have seen how some mathematicians, including Descartes, did not countenance negative numbers as coefficients or solutions, because how can you have a negative length? Or a negative area? But now with a pair of coordinate axes, a negative x value for example is simply on the left of the y-axis, rather than the right. And what numbers are on the axes themselves? Certainly the natural numbers $(0, 1, 2, 3, \ldots$ which we denote $\mathbb{N}$) but now if we also accept negative numbers we must extend to the integers $(0, \pm 1, \pm 2, \ldots$ which we denote $\mathbb{Z}$ from the German *zahlen* for 'number'). In the space between the integers, we should also include rational numbers, those formed by a ratio $\frac{a}{b}$ where a and b are integers (and $b \neq 0$); these are denoted $\mathbb{Q}$, for quotient.

We could imagine it in the following way: draw a pair of axes, label them x and y. Now draw vertical lines through $x = 1, -2, 3, \ldots$; every point on those lines has that x coordinate. Now draw horizontal lines through $y = 11, -5, 38, \ldots$; every point on those lines has that y coordinate. Let's keep going: draw vertical lines through $x = \frac{1}{2}, \frac{-7}{3}, \frac{21}{37}, \ldots$ and all the rational numbers, then horizontal lines through $y = \frac{-2}{17}, \frac{54}{9}, \frac{-113}{26}, \ldots$ and all the rational numbers; you can see we are laying down an infinitely fine mesh of horizontal and vertical lines. Where one of the vertical and

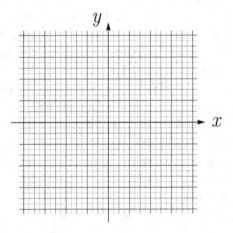

horizontal lines meet then both the x and the y coordinates of that point have rational values, and so we call these 'rational points', and we have covered the x-y plane in an infinitely fine dust of rational points, as close together as you like. The question is: is that it? Are we done?

Consider the equation

$$x^3 + y^3 = 1.$$

This is an equation in x and y, and so it represents a curve; we draw that curve in the figure below:

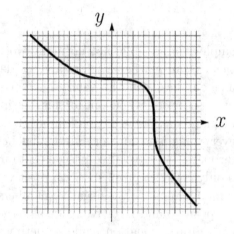

Now here's the thing: we mentioned Fermat's Last Theorem previously, and this is a special case: this equation has no rational solutions (other than when x or y are zero). In other words, there are no pairs of rational numbers x and y which satisfy this equation; or to put it yet another way, the curve defined by this equation *does not pass through any rational points in the plane*. So while we have defined an infinitely fine dust of rational points, this curve manages to snake its way across the plane without passing through any of them! Pythagoras would be rolling in his grave. This curve somehow finds the gaps in the rational numbers, so we must extend our notion of number to include irrational numbers, and we denote this extension with $\mathbb{R}$, the real numbers; already in 1585 Stevin was thinking of the number line as a continuum [119]. Algebra was steaming ahead, everything seemed to be falling into place and the future was bright; the only niggle was that sometimes what's under the square root sign is negative, but that can't be that big a deal, can it?

CHAPTER 6

Complex Numbers

I RRATIONAL, complex, impossible; not normally words you associate with mathematicians, but mathematicians are people too and we can be prone to putting our heads in the sand. There is a famous story that Pythagoras was so upset when one of his disciples proved that the diagonal of a square was not a rational multiple of the side, thus confounding his 'everything is ratio' worldview, that he had his hapless protégé thrown overboard. We have already seen that our ancestors were perfectly happy to only consider problems with positive coefficients and solutions, for example Cardano called negative numbers 'fictitious' and Descartes described positive and negative roots as 'true' and 'false' respectively. That a quadratic equation could sometimes have solutions and sometimes not was not a difficult pill to swallow: if I asked you for the points of intersection between a line and circle you would have no problem saying sometimes there are two points of intersection and sometimes there are none (like on the left below).

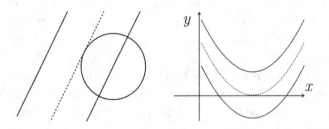

Figure 6.1: Roots or no roots?

A more Cartesian way to phrase this would be to say if $y = ax^2 + bx + c$, a quadratic in x, which Descartes and Fermat knew was a parabola in the

DOI: 10.1201/9781003455592-6

plane (like on the right in Figure 6.1), then the roots of $ax^2 + bx + c = 0$ are simply where the parabola crosses the x-axis, and that might happen twice or not at all.

This changed with cubics. Suppose we want to solve

$$x^3 - 15x = 4.$$

We could use Viète's formula given in (5.2) with $B = 5$ and $Z = 2$ to get (using $Z^2 - B^3 = -121$) the root

$$\sqrt[3]{2 + \sqrt{-121}} + \sqrt[3]{2 - \sqrt{-121}}, \qquad (6.1)$$

but now we can't just wave this away and say 'oh we will ignore this case as we only want to consider cubics with real roots', because *all* cubics must have a real root. We can see this is true by the following argument: if $y = $ a cubic, say something like $y = x^3 + \ldots$, then since this is a curve in the x-y plane we can see that as x gets very large and positive then y is also getting large and positive, whereas as x gets large and negative then so does y; a picture of the graph of y, just with these limiting parts drawn in, looks like this:

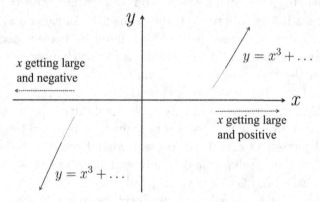

But the curve has to cross the x axis to get from one side to the other, so there must be a real root somewhere on the way. This is a great example of switching back and forth between seeing a problem geometrically and seeing it algebraicly, but also note we are assuming that the curve cannot 'jump' from one side of the x-axis to the other; we are thinking of the curve as 'continuous', and this sort of thinking will be common when we come to the Topology part of this book.

So $x^3 - 15x = 4$ must have a real root (in fact all the roots of this particular cubic are real), which means we must take things like $\sqrt{-121}$

seriously. Cardano was aware of this, and in his *Ars Magna* you can see the first acknowledgement of the square root of a negative number, 'R̃ m 15' being $\sqrt{-15}$ (Figure 5.4 on the right), but he dismissed this as *usere et sophistica*, useless and 'sophistic' (refined/subtle). In 1572, Rafael Bombelli in his *Algebra* [68] attacked the issue head on: he realized that the complicated expression in (6.1) was simply the number 4 in disguise, and he set about deriving rules for how to work with sophisitic numbers. For example, he called $\sqrt{-1}$ *più di meno* and $-\sqrt{-1}$ *meno di meno*, and said [119] *meno di meno uia meno di meno fà meno*, or $(-\sqrt{-1}) \times (-\sqrt{-1}) = -1$.

Despite this advance the idea of working with these numbers was very slow to catch on; Leibniz referred to them as 'amphibian' [16] and Newton called them 'impossible' [109], but gradually the term 'imaginary' for multiples of $\sqrt{-1}$ came to be used, with combinations of the form $a + b\sqrt{-1}$ (where a and b are real numbers) called 'complex'. I think it is fair to say that the labels 'imaginary' and 'complex' have been putting people off this topic for centuries; like Led Zeppelin and Daft Punk, 'imaginary' was originally a derogatory term that stuck, used by Descartes in *La géométrie*, and 'complex' is used in the sense of several things joined together, like a leisure complex, rather than the sense of being difficult. It was only really through the work of Euler and Gauss in the 18th and 19th centuries that complex numbers became accepted by the community at large, and so I will focus in this chapter on two major results from the theory of complex numbers, one due to Euler (Euler's formula) and one due to Gauss (the fundamental theorem of Algebra).

In keeping with the tone of this book, I will give a historical sketch of these two results but try to avoid as much as possible the technical details and jargon; as such I will present what I consider to be elegant and accessible proofs relying on diagrams and ideas, but these are not the proofs that Euler and Gauss gave originally (although they are related). As a warm up, let us develop a more intuitive and geometrical way to view complex numbers, which we touched on in Chapter 4.

The 'traditional' approach to complex numbers is via their arithmetic rules: how to add, multiply, divide and so on, just like Bombelli did. Euler had introduced the symbol i for the imaginary unit $\sqrt{-1}$ in the 1770s, so we will use it here. Suppose $a + bi$ and $c + di$ are complex numbers, with a, b, c, d all real numbers, then we can add complex numbers like so:

$$(a + bi) + (c + di) = (a + b) + (c + d)i$$

and we can multiply like so:

$$(a + bi) \times (c + di) = (ac - bd) + (ad + bc)i, \qquad (6.2)$$

but this is a bit dry and formal. In 1797 Wessel [68] wanted to devise a way for a number to capture both a magnitude *and* a direction; he realized that if you take a complex number like $a + bi$, and we separate a (the 'real' part) and b (the 'imaginary' part), then we can plot the point (a, b) in the plane, and now $a + bi$ describes the line segment from the origin to the point (a, b), which has both a magnitude and a direction (the arrow emphasizing we are going *from* the origin *to* the point (a, b)).

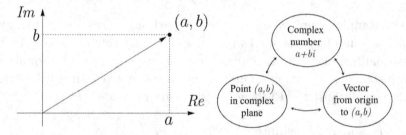

We have labeled the horizontal axis Re, since this is for the real part, and the vertical axis Im since this is for the imaginary part, and together they span the 'complex plane' (although Wessel's paper was largely unnoticed and the idea was independently put forward by Argand in 1806, so we sometimes call this the 'Argand plane'). Now $a + bi$ can be viewed as the point in the complex plane with coordinates (a, b), or we can also think of it as the directed arrow that goes from the origin to the point (a, b). This 'directed arrow' is an early instance of what came to be known as a 'vector', although that term was only introduced by Hamilton about 50 years later (see Chapter 8).

This is a wonderful step forward, because complex numbers were abstract and mysterious, but points in the plane are intuitive and well understood. This is the reason some would argue that complex numbers are just as real as real numbers, the only difference is that they are two-dimensional. We can now understand addition and multiplication in geometrical terms: $a + bi$ and $c + di$ are vectors that point from the origin to (a, b) and (c, d) respectively, which form the sides of a parallelogram; their sum is simply the diagonal of that parallelogram, or $(a, b) + (c, d) = (a + c, b + d)$ as we wrote it in Chapter 4.

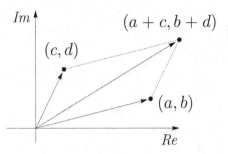

We can also see where the multiplication rule from Chapter 4 came from, just by writing (6.2) in coordinates:

$$(a, b) \times (c, d) = (ac - bd, ad + bc).$$

Every complex number has a twin or partner, called the 'conjugate', which is found by replacing i with $-i$ (so the conjugate of $3+4i$ is $3-4i$). With multiplication and conjugate defined we can derive a formula for the distance of a complex number, usually denoted z, from the origin (this is really just the Pythagoras theorem in disguise):

$$\left(\begin{array}{c} \text{square of distance} \\ \text{of } z \text{ from the origin} \end{array} \right) = \left(\; z \; \right) \times \left(\; \text{its conjugate} \; \right). \qquad (6.3)$$

We are now ready to derive the most beautiful equation in Mathematics.

In 1748 Euler published *Introductio in analysin infinitorum*, Introduction to the Analysis of the Infinite [38]. In it, Euler provides an extensive description of power series, highlights the key role of e^x, makes central the notion of function as well as introduces the $f(x)$ notation, and establishes complex numbers as a vital and unavoidable part of mathematical analysis; all of these ingredients we will explain and use below. Tucked away in the *Introductio*, in Section 138 (see Figure 6.2), is an expression that has come to be known as 'Euler's Formula' which we will derive (although the famously prolific Euler has several formulae named after him, we will see another in Chapter 13). Euler intended the *Introductio* as a precursor to his later texts on Calculus, so we will follow him and work without the tools of Calculus.

Let us begin with polynomials, expressions like $y =$ some finite sum of positive powers of x. We have already seen quadratics $y = a_0 + a_1 x + a_2 x^2$

pra invento, perpetuo $e^z = 1 + \frac{z}{1} + \frac{z^2}{1.2} + \frac{z^3}{1.2.3} + \frac{z^4}{1.2.3.4} + \&c.$
ipfi vero Logarithmi hyperbolici ex his Seriebus invenientur ,.

Ex quibus intelligitur quomodo quantitates exponentiales ima-
ginariæ ad Sinus & Cofinus Arcuum realium reducantur. Erit
vero $e^{+v\sqrt{-1}} = cof. v + \sqrt{-1}. fin. v$ & $e^{-v\sqrt{-1}} = cof. v - \sqrt{-1}. fin. v.$

Figure 6.2: Excerpts from Euler's 1748 *Introductio in analysin infinitorum* [38], Sections 123 and 138. Note Euler is still writing $\sqrt{-1}$, only
introducing the i notation about 30 years later.

and cubics $y = a_0 + a_1 x + a_2 x^2 + a_3 x^3$ but we can keep going, in fact we
can keep going forever:

$$y = a_0 + a_1 x + a_2 x^2 + a_3 x^3 + a_4 x^4 + a_5 x^5 + \ldots$$

where by ... we mean this expression just keeps going and going to include all positive integer powers of x; as such this is called an 'infinite
series' or 'power series'. Note the notation: x for the unknown and a for
the coefficients (just like Descartes), and the subscript on the a matches
the power on the x. What distinguishes one power series from another
is the choice of coefficients, for example this is one power series

$$1 + x + x^2 + x^3 + x^4 + \ldots$$

and this is another

$$1 - \frac{x^2}{2} + \frac{x^4}{2.3.4} - \frac{x^6}{2.3.4.5.6} + \ldots$$

but arguably the most important example of a power series is

$$1 + \frac{x}{1} + \frac{x^2}{1.2} + \frac{x^3}{1.2.3} + \frac{x^4}{1.2.3.4} + \ldots$$

as Euler wrote it in Section 123 of the *Introductio* (see Figure 6.2). This
series is called the 'exponential function', $\exp(x)$, since it has some of
the properties of exponentiation, for example

$$\exp(x_1 + x_2) = \exp(x_1)\exp(x_2). \tag{6.4}$$

(This function also has the distinguishing property that it is equal to its own derivative, which we will say more of in Chapter 9).

To evaluate this function for some x we just sub in that value for x, for example $\exp(0) = 1$ and $\exp(1)$ is

$$1 + \frac{1}{1} + \frac{1}{1.2} + \frac{1}{1.2.3} + \frac{1}{1.2.3.4} + \cdots$$

which evaluates to 2.718... Already in the 1720s Euler was using the letter e to represent this number, and it has been customary to write $\exp(x)$ as e^x; so for example when we write e^π we mean $\exp(\pi)$ or

$$1 + \frac{\pi}{1} + \frac{\pi^2}{1.2} + \frac{\pi^3}{1.2.3} + \frac{\pi^4}{1.2.3.4} + \cdots$$

which evaluates to about 23.1407... It is important to stress that e^π does not mean 'the number 2.718... multiplied by itself 3.14159... times', like 2^3 means 'the number 2 multiplied by itself 3 times'.

Now we can ask, what if we let x be a complex number? For example if $x = it$, where t is some real number, what is e^{it}? The first thing to point out is that since we have an i in there, then we might suppose e^{it} is a complex number (although there can be surprises: for example i^i is real!). As such, for any given value of t, then e^{it} is a point in the complex plane. Choosing different values for t will give different points, and indeed letting t vary continuously will sweep out a locus of points in the plane, in other words:

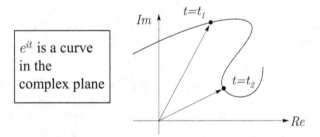

And what curve is it? Suppose we want to know the distance of a point on this curve from the origin. To use (6.3) we need the conjugate of e^{it} which is, after replacing i with $-i$, just e^{-it}. Now we have

$$\left(\begin{array}{c} \text{square of distance} \\ \text{from the origin} \end{array} \right) = e^{it} \times e^{-it} = \left(\text{using equation (6.4)} \right) = e^0 = 1.$$

So e^{it} is the curve with the property that every point on it is a distance 1 from the origin, but that curve is just the unit circle. Since e^{it} is a complex number it will have a real part (the x-axis) and an imaginary part (the y-axis), and we already saw in (3.1) how to write the x and y coordinates of points on a circle as functions; these are then the real and imaginary parts of e^{it} and we arrive at Euler's formula (Figure 6.2):

$$e^{it} = \cos(t) + i\sin(t).$$

Different values of t will pick out different points on the circle; in particular, $t = \pi$ gives the left-most point with coordinates $(-1, 0)$ which is the complex number -1, so

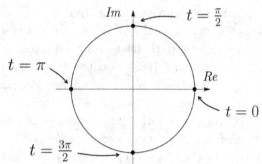

$$e^{i\pi} = -1.$$

This is often voted 'the most beautiful equation in mathematics', but I always found the -1 a bit incongruous; Boyer [16] prefers to write it like this:

$$e^{i\pi} + 1 = 0,$$

a single equation containing e, i, π (the three most fundamental constants in mathematics), 1 and 0 (the fundamental numbers of mathematics), together with $+$ and $=$ (the most fundamental operation and relation in mathematics), all in one place.

I am using the word 'fundamental' a lot, but it is hard not to: we are now going to prove the Fundamental Theorem of Algebra, perhaps first stated by Girard [68] in 1629 as *every algebraic equation admits of as many solutions as the denomination of the highest quantity indicates*. We would phrase this today as: a polynomial of degree n has n roots, if we allow the roots to be complex and we count roots by their multiplicity. For example if you look back to Figure 6.1 the theorem asserts that there are always two roots to a quadratic even when the parabola doesn't cross the x-axis, it's just that they will then be complex; moreover when the parabola is just touching the axis then there is a 'double root', a root we count twice. Girard did not give a proof, d'Alembert in 1746 was incomplete, and even Euler in a 1751 attempt didn't manage it [114]. Gauss was a little obsessed with this theorem, and in fact gave four proofs in his lifetime [68], the first being his 1799 doctoral thesis, however he is generally not credited with the first proof of the theorem. That honour goes to Argand in 1814, the reason being that Gauss's proof had a gap (even a giant can stumble, but then, he was only 22. The explanation of this gap will make more sense after our own version below). In fact Gauss and others phrased the fundamental theorem slightly differently: any polynomial of degree n, even one with complex coefficients, must have at least one complex root. If that is true, we can divide out by a factor to get a polynomial one degree less, and then by the same argument this remaining part must also have a root, and so on. We will prove the original phrasing (based on the excellent *Mathematical Omnibus* [43]), using only careful diagrams, clever ideas and thinking more precisely about what we mean when we say a function is 'continuous'.

By the early 19th century mathematicians were stressing the importance of 'rigor', proving something beyond a doubt with no wriggle room; gone were the days where Descartes could make a claim and say *I have intentionally omitted [explanation] so as to leave to others the pleasure of discovery* [31]. An early advocate was Augustin-Louis Cauchy, who in 1821 [68] gave a formal definition of what it means for a function to be continuous, which we will adapt in the following way: suppose your function is a polynomial of degree n, let's call it $f(z)$ (I am using z rather than x as we are thinking more generally in terms of complex numbers):

$$f(z) = a_0 + a_1 z + \ldots + a_{n-1}z^{n-1} + a_n z^n,$$

with neither a_n nor a_0 equal to zero. We can think of it like a machine, with an input (a complex number) and an output (a complex number). Picturing complex numbers as points in the plane, we can see it like this:

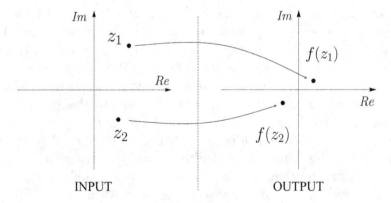

If you input the number z_1, the function outputs $f(z_1)$, and I like to use the phrase 'gets mapped to', as in 'the number z_1 gets mapped to $f(z_1)$'. In particular, if z_1 gets mapped to the origin, then z_1 must be a root of f, because $f(z_1) = 0$. Suppose there are two inputs, z_1 and z_2, with corresponding outputs $f(z_1)$ and $f(z_2)$. When we say 'the function f is continuous' we mean that if z_1 and z_2 are close, then $f(z_1)$ and $f(z_2)$ are also close. How close? As close as you like!

But now rather than just a point on the left getting mapped to a point on the right, let's draw a curve on the left and, taking each point on the curve and putting it into f, we get another curve on the right (see Figure 6.3). We could say that the curve γ gets mapped to another curve, $f(\gamma)$. I know $f(\gamma)$ must be a curve, rather than some disjoint set of points, because f is continuous: two nearby points on γ must get mapped to two nearby points on $f(\gamma)$. Moreover, if γ is 'closed', by which I mean it joins up with itself and doesn't have loose ends, then so must be $f(\gamma)$, again because f is continuous.

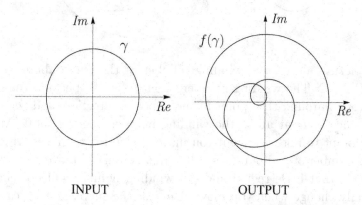

INPUT   OUTPUT

Figure 6.3: The function f maps the 'input' curve on the left, γ, to the 'output' curve on the right, $f(\gamma)$.

For the image above I let $f(z)$ be a cubic, and the circle on the left was then mapped to the curve on the right. Note the curve on the right goes through the origin; this means the circle on the left must pass through a root of $f(z)$, because some point on that circle got mapped to the origin.

The resulting curves can be all loopy and twisty, and it is this loopiness we are going to make more precise by defining the 'winding number' (a notion also due to Cauchy [90]; we will see winding numbers again in Chapter 14 as one instance of a much bigger idea). It goes like this: suppose you have a closed curve in the plane, and you give it an 'orientation', a sense of direction which we indicate with arrows. Now pick some point in the plane not on the curve, let's call it A. Suppose you are standing at A watching a point make its way around the curve following the orientation; when it has performed a complete circuit, you will have turned around anti-clockwise a certain number of times. This is called the 'winding number' of the curve about A, because it captures how many times the curve winds around that point (if the winding number is negative it just means there is a total *clockwise* winding).

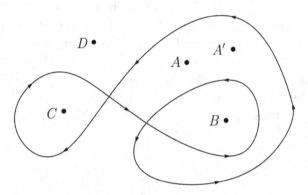

In the picture above, the winding number of the curve about A is 1, about B it's 2. The winding number about C is -1, because the curve has wrapped around that point in the opposite direction, and the curve doesn't enclose D at all so the winding number about D is 0. An important point to note is that if you move A slightly, to A' say, then the winding number of γ about A' is still 1. In fact you can move A anywhere you like in that little region and the winding number will not change; it will only change when you move A *across* the curve. But we can also think of moving or deforming the curve itself; again small changes to the curve γ will not change the winding number of γ about A, until the curve is dragged *across* A; then the winding number changes. We are now ready to prove the fundamental theorem of algebra.

Let's look back at the 'input' curve γ being a circle, Figure 6.3, and think about the winding number of the 'output' curve $f(\gamma)$ about the origin. Imagine making the input circle larger and smaller, gradually and continuously; the output curve will deform, also gradually and continuously, no 'jumps'. Let's look at two extreme cases: suppose the input circle is very very small, then z is also very very 'small', which means in our polynomial

$$f(z) = a_0 + a_1 z + \ldots + a_{n-1} z^{n-1} + a_n z^n$$

all the terms with powers of z are getting so small that the constant term, a_0, is dominant. So this teeny tiny input circle gets mapped to a curve which only encloses a_0 and a small region around it (left-most in Figure 6.4); therefore the winding number of this curve about the origin is zero.

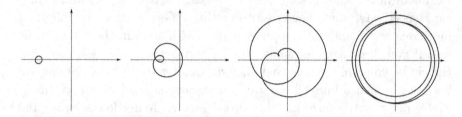

Figure 6.4: The output curve deforming as the input circle goes from very small (left) to very large (right); the winding numbers about the origin are $0, 1, 2$ and 3.

Now let's look at the other extreme: the input circle is very very large. In our polynomial, the highest term, z^n, is now dominant, so much so that $f(z)$ becomes more and more like just $f(z) = a_n z^n$, in which case the output curve wraps around the origin n times (right-most in Figure 6.4). Finally imagine continuously going from one extreme to the other, from a very small input circle to a very large one. The output curve must go from having winding number about the origin 0 to having winding number n, but this is only possible if the curve crosses over the origin n times. Every time it passes through the origin that means there was a point on the input circle which got mapped to the origin which is therefore a root; since it passes over the origin n times there must be n roots.

It is said [43] that *nearly every branch of mathematics tests its techniques and proves its maturity by providing a proof of the fundamental theorem of algebra*, and there are indeed very many proofs of it, so where did Gauss go wrong? His approach was as follows: start with a polynomial in z like we have above, and sub in $z = a + bi$. When you multiply everything out, you will have

$$f(a + bi) = \underbrace{(\text{expression with } a \text{ and } b)}_{=u} + i\underbrace{(\text{expression with } a \text{ and } b)}_{=v}.$$

To prove a root must exist, we just need to prove that the two expressions labeled u and v must both vanish simultaneously at least once; or to put it another way, if you draw the curves in the a, b plane where $u = 0$, and the curves where $v = 0$, then they *must* cross one another somewhere. For example, using the same cubic as in the images in the last pages, I draw in heavy and dashed lines below the curves where u and v vanish respectively; note they cross three times. Gauss made the following ingenious argument (no computer generated images in those days!): he said if you draw a large enough circle, and note where the curves cut the circle, you will observe that they do so alternately: a heavy line followed by a dashed line, all the way round; how elegant! Based on this he said it must be the case that they cross somewhere inside the circle, and therefore there is a root; while this seems reasonable crucially he didn't prove this statement, saying [109] *if anybody desires it, then on another occasion I intend to give a demonstration which will leave no doubt.*

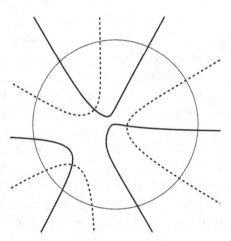

It is often said that Topology started with Euler and the bridges of Königsberg, and indeed that is where Part IV of this book will begin, but in my opinion *this* is where Topology started in earnest: in complex analysis, and attempts to make intuitive notions about deforming curves rigorous; we will investigate this in more depth in Part IV.

So now we know that a polynomial of degree n has n roots, how do we actually calculate them? We just use the formula, right?

Abstract Algebra

A T THE START of the 1800s, Algebra was about solving equations. By the end of the 1800s, Algebra had developed into an abstract and powerful meditation on structure, form and relation. I have heard it described in the following way: we build a tower, and as we climb the tower we see less and less of the detail of the ground nearby, but in return we start making out the broader features of the surrounding landscape. The higher we go, the more we can appreciate the common structures of the world around us, free to see the similarities without getting bogged down in superfluous details. But not everyone has a head for heights, and in this chapter I hope to give a hint of the view from the tower to see if it is to your taste.

Our first step on the winding stair is to think more carefully about what we mean when we say 'solve an equation' or 'find the roots of a polynomial'. Consider the general quadratic equation:

$$ax^2 + bx + c = 0$$

where we will suppose in what follows that a, b, c are unspecified rational numbers (i.e. ratios of integers), and more generally whenever I refer to an equation we will suppose it to have rational coefficients. As Brahmagupta knew, but expressed in words, the roots of this equation are

$$\frac{-b + \sqrt{b^2 - 4ac}}{2a} \quad \text{and} \quad \frac{-b - \sqrt{b^2 - 4ac}}{2a}, \tag{7.1}$$

but let's pause over this for a moment. What we are saying is: we can construct the roots by taking the coefficients in the equation, a, b, c, and other rational numbers such as -4 and 2, and combine them using the

DOI: 10.1201/9781003455592-7

elementary operations of addition and subtraction, multiplication and division, as well as taking roots. If the roots (unfortunate that the word 'root' has two subtly different meanings here) to an equation can be constructed in this way, we say the equation is 'solvable by radicals' or just 'solvable'; so all quadratic equations are solvable. As we saw in Chapter 5, by the 1500s Cardano and Viète knew that the cubic equation was solvable by radicals, and in Cardano's *Ars Magna* he gave the solution to the general quartic equation, due to his colleague Ludovico Ferrari. So already by 1600 we knew that all equations of degree 1,2,3 and 4 are solvable by radicals, and we might naturally assume this is the case for all polynomial equations of arbitrary degree. In particular, the search was on to find the formula for the roots of the general quintic equation,

$$ax^5 + bx^4 + cx^3 + dx^2 + ex + f = 0,$$

by which we mean: can we make (possibly very elaborate) combinations of the coefficients of this equation and other rational numbers using the elementary operations of arithmetic and extracting roots in order to construct the roots of the equation? The surprising answer to this question is: no. In 1799 Ruffini [16] published the first proof that not all polynomials of degree greater than 4 are solvable by radicals, i.e. there is no 'quintic formula', although some found his proof (as well as his earlier attempts) to be incomplete; in 1824 Abel [109] published the first complete proof that not all quintics are solvable by radicals, the classic example of such an unsolvable quintic being

$$x^5 - x + 1 = 0. \tag{7.2}$$

This equation has 5 roots as we know, however they cannot be constructed by combining rational numbers, elementary operations, and extracting roots, no matter how much you try; even if you cook up a beast like

$$\frac{\sqrt[3]{321 - \sqrt{-122 + \sqrt[4]{33} + \frac{2122}{\sqrt[7]{221}}}}}{\sqrt{2 - \frac{11}{\sqrt[3]{-22556}}}},$$

and many others besides, and make many combinations of them and take more roots and more combinations and so on until the cows come home, you will still not have reached the roots of (7.2). On the other hand, this quintic

$$x^5 + x^4 + 1 = 0$$

is solvable by radicals. Why are some quintics solvable, and some are not? What is special about going from degree 4 to degree 5? To address these much deeper questions a whole new framework was needed, which turned out to have implications for all of Algebra and indeed all of Mathematics, and it came from one of the great tragic heroes of the 19th century: Évariste Galois.

It would be fair to say that Galois had a problem with authority. Mathematicians often began their books and memoirs by carefully thanking their patrons and giving credit to their predecessors; not so Galois. He writes [41] *this work is not encumbered with names, forenames, qualities, titles, and elegies of some miserly prince whose purse will be opened with the fumes of incense - with the threat of being closed when the censer-bearer is empty... If I had to address anything to the great in the world or the great in science... I swear that it would not be thanks. I owe to the ones... that I have published so late, to the others that I have written it all in prison.* What a legend!

Born in Paris in 1811, Galois was home-schooled and only knew the rudiments of arithmetic when he was 12, but by the time he was 17 he was having original research published in respected journals [119]. In 1831, aged just 19, Galois submitted to the Académie des Sciences de Paris his *Mémoire sur les conditions de résolubilité des équations par radicaux* in which, in a startling display of vision and originality, Galois laid the foundations of what has come to be known as 'group theory' and applied it to the question of solvability of polynomial equations (now known as Galois theory), culminating in a criterion for determining which equations are solvable by radicals and which are not. Less than two years later, Galois was shot dead. I will first give an idea of what Galois had achieved in his famous *mémoire* before returning to the sad story of how this brilliant but hot-headed young man threw his life away.

Let us begin with the cubic equation $x^3 + ax^2 + bx + c = 0$, and suppose it has three distinct roots A, B, C. As such we can factor:

$$x^3 + ax^2 + bx + c = (x - A)(x - B)(x - C)$$

and, multiplying out the brackets on the right hand side and equating coefficients of the various powers of x we get three equations (these were already known to Viète and Girard):

$$A + B + C = -a, \quad AB + AC + BC = b, \quad ABC = -c. \qquad (7.3)$$

The thing to notice here is that if we swap around some of the roots, say swap A and B, then these three equations 'still hold', by which I mean: if $A+B+C$ is equal to $-a$ then $B+A+C$ will still be equal to $-a$ (after all, who is to say which root we should call A and which we should call B?); these equations are *symmetric* in this sense. In fact if you perform *any* rearrangement of the roots A, B, C these three equations still hold, and the more general question is: can we say the same for all polynomial equations in A, B, C? The answer to that tells us something about the nature of the roots and hence leads to information about whether the equation is solvable or not. This focus then on the set of all possible rearrangements, or 'permutations', of the roots lead to the introduction of the notion of a 'group', as we will see.

Galois was not the first to think about permuting the roots in this way; already Vandermonde, Lagrange and Cauchy contemplated symmetric polynomials [119], and Cauchy introduced a notational device for permutations which we adapt here:

$$\begin{pmatrix} 1 & 2 & 3 & 4 \\ 2 & 1 & 3 & 4 \end{pmatrix}$$

means 'in the list of four symbols, swap the first and second elements', so $ABCD$ goes to $BACD$. Galois noted that we can compose permutations, by just doing one after the other, and the result is itself a permutation, or to put it another way: composing any two elements of the set of permutations gives you an element of the set of permutations. For example, consider the two following permutations:

$$T = \begin{pmatrix} 1 & 2 & 3 & 4 \\ 2 & 1 & 3 & 4 \end{pmatrix} \quad \text{and} \quad S = \begin{pmatrix} 1 & 2 & 3 & 4 \\ 1 & 3 & 2 & 4 \end{pmatrix}.$$

If we start with $ABCD$ and perform the permutation T ('swap the first two elements') we get $BACD$, and if we do to *that* the permutation S ('swap the middle two elements'), we get $BCAD$, but this is the same as the permutation

$$\begin{pmatrix} 1 & 2 & 3 & 4 \\ 3 & 1 & 2 & 4 \end{pmatrix}$$

(read it vertically: the first column $\frac{1}{3}$ means 'the symbol that was in the first slot is sent to the third slot', the symbol that was in the second slot is sent to the first slot, and so on). We will write this as $S \circ T$, which you can read as 'S composed with T' or 'S after T'. It is a good time to point out that if we had done S first and then T, we would get a different result ($CABD$), or in other words, $S \circ T$ is not the same as $T \circ S$; order matters! Galois called this the 'group of permutations', but the problem was he didn't say exactly what he meant. Indeed over the next decades other mathematicians worked with groups but the modern definition was some time coming, mostly because they were originally focused on just permutations which automatically have certain properties they felt didn't need to be explicitly stated. Probably the first to come close to the modern definition was Weber (1882), and we give here the currently accepted definition which you will find in a modern text (Lang [72] is the standard reference, but Jordan and Jordan [65] is more approachable):

Suppose you have a set of 'things' (such as the set of all permutations of the list of n symbols), and some operation $\circ$ you can perform on a pair of elements of that set (such as composing permutations). The set, with this operation, is a *group* if the following four properties hold:

1. The set is closed (if S and T are in the set then $S \circ T$ is also in the set),

2. There exists an identity element (there is some element E such that $S \circ E$ and $E \circ S$ is equal to S, for any S in the set),

3. Every element has an inverse (for any S in the set there is some other element T in the set such that $S \circ T$ is the identity, E),

4. The operation is associative (for any elements S, T, U of the set we must have $S \circ (T \circ U)$ is equal to $(S \circ T) \circ U$, where the brackets indicate which to perform first).

Let's check how this works for the set of permutations of the list of three symbols, which we can denote S_3. How many elements does this set have? Well after a permutation there are three slots to fill, with three choices for the first slot (A, B or C), two for the second and one for the last (you might have come across the word 'permutation' in this combinatorial context before), so the six elements of S_3 are

$$E = \begin{pmatrix} 1 & 2 & 3 \\ 1 & 2 & 3 \end{pmatrix}, \quad T_1 = \begin{pmatrix} 1 & 2 & 3 \\ 2 & 1 & 3 \end{pmatrix} \quad T_2 = \begin{pmatrix} 1 & 2 & 3 \\ 3 & 1 & 2 \end{pmatrix}$$

$$T_3 = \begin{pmatrix} 1 & 2 & 3 \\ 1 & 3 & 2 \end{pmatrix}, \quad T_4 = \begin{pmatrix} 1 & 2 & 3 \\ 2 & 3 & 1 \end{pmatrix} \quad T_5 = \begin{pmatrix} 1 & 2 & 3 \\ 3 & 2 & 1 \end{pmatrix}$$

If you compose any two of these you will get another one, for example composing T_2 with T_5 gives T_1 (or $T_2 \circ T_5 = T_1$), and composing T_5 with itself gives E. What's more E is clearly the identity element ('do nothing'), and every element has an inverse, for example composing T_2 with T_4 gives the identity E, so T_2 and T_4 are inverses of each other (the last property, associativity, can be tedious to check so we will overlook it in what follows). So S_3, the set of permutations of the list of 3 symbols, with the operation of composition, is a group. In general the set of permutations of the list of n symbols, denoted S_n, with the operation of composition, is a group (with $n!$ elements).

Now notice that in S_3 the element T_5 when composed with itself gave the identity, so T_5 is its own inverse. That means that this set:

$$E = \begin{pmatrix} 1 & 2 & 3 \\ 1 & 2 & 3 \end{pmatrix} \quad T_5 = \begin{pmatrix} 1 & 2 & 3 \\ 3 & 2 & 1 \end{pmatrix}$$

is itself a group, since it satisfies all the required properties. This group, being a subset of S_3, is called a *subgroup* of S_3, and in general, a group can contain many subgroups within it, however the number of elements in each subgroup must divide the number of elements in the parent group (for example this subgroup of S_3 has two elements, and S_3 has six elements, and 2 divides 6; in particular S_3 cannot have a subgroup with 4 elements). This is called Lagrange's theorem, based on work of Lagrange over 100 years before there was a formal definition of a group... but let's go back to our cubic equation.

The three roots of a cubic A, B, C form a list of 3 symbols and therefore there are $3! = 6$ possible rearrangements of them, the set S_3. We can see that each of the three symmetric polynomials given in (7.3) still hold after any one of these rearrangements, or any element of S_3. Can we say the same for any polynomial equation in A, B, C? Suppose the cubic is $x^3 - 2x^2 - 2x + 4 = 0$ and the roots are known explicitly: $A = 2, B = \sqrt{2}, C = -\sqrt{2}$. Equations like

$$A^2 + B^2 + C^2 = 8, \quad A^2 B^2 C^2 - 3(A + B + C) = 10$$

are still true after any permutation of the roots, however equations like

$$A - 2 = 0, \quad A + C^2 = 4, \quad B^3 + AB^2 - 2B = 4$$

are not still true after *any* permutation in S_3. However, they are still true after the permutation $\left(\begin{smallmatrix} 1 & 2 & 3 \\ 1 & 3 & 2 \end{smallmatrix}\right)$, and in fact, any polynomial equation stays true after $\left(\begin{smallmatrix} 1 & 2 & 3 \\ 1 & 3 & 2 \end{smallmatrix}\right)$ (just try it! Any equation you write down which is satisfied by $2, \sqrt{2}$ and $-\sqrt{2}$ must still be satisfied if you replace $\sqrt{2}$ with $-\sqrt{2}$, if the equation has rational coefficients). The same can be said for the identity permutation. Thus there is a subset of S_3, which is the set $\{(\begin{smallmatrix} 1 & 2 & 3 \\ 1 & 2 & 3 \end{smallmatrix}), (\begin{smallmatrix} 1 & 2 & 3 \\ 1 & 3 & 2 \end{smallmatrix})\}$, such that any equation in A, B, C which is true before a permutation in this set will still be true after that permutation (you could say that polynomial equations cannot distinguish these two roots). With composition as an operation, this set is itself a group, and therefore a subgroup of S_3. We can now make an important definition: for a polynomial of degree n, the subgroup of S_n for which any polynomial equation in the roots still holds after those permutations is called the *Galois group* of that polynomial; the Galois group of $x^3 - 2x^2 - 2x + 4 = 0$ is therefore $\{(\begin{smallmatrix} 1 & 2 & 3 \\ 1 & 2 & 3 \end{smallmatrix}), (\begin{smallmatrix} 1 & 2 & 3 \\ 1 & 3 & 2 \end{smallmatrix})\}$.

Another example: the equation $x^4 - 2x^2 - 3 = 0$ has roots

$$A = \sqrt{3}, \quad B = -\sqrt{3}, \quad C = \sqrt{-1}, \quad D = -\sqrt{-1}.$$

Clearly not all elements of S_4 will leave every polynomial equation true, for example $A^2 - C^2 = 4$ will not hold if we perform $(\begin{smallmatrix} 1 & 2 & 3 & 4 \\ 3 & 2 & 1 & 4 \end{smallmatrix})$, since $C^2 - A^2$ is not equal to 4. However every equation *will* still hold after we perform $(\begin{smallmatrix} 1 & 2 & 3 & 4 \\ 2 & 1 & 3 & 4 \end{smallmatrix})$ or $(\begin{smallmatrix} 1 & 2 & 3 & 4 \\ 1 & 2 & 4 & 3 \end{smallmatrix})$, or indeed both at the same time: $(\begin{smallmatrix} 1 & 2 & 3 & 4 \\ 2 & 1 & 4 & 3 \end{smallmatrix})$. This means the set

$$\left\{ \begin{pmatrix} 1 & 2 & 3 & 4 \\ 1 & 2 & 3 & 4 \end{pmatrix}, \begin{pmatrix} 1 & 2 & 3 & 4 \\ 2 & 1 & 3 & 4 \end{pmatrix}, \begin{pmatrix} 1 & 2 & 3 & 4 \\ 1 & 2 & 4 & 3 \end{pmatrix}, \begin{pmatrix} 1 & 2 & 3 & 4 \\ 2 & 1 & 4 & 3 \end{pmatrix} \right\}, \quad (7.4)$$

a subgroup of S_4, is the Galois group of the polynomial.

In some sense the Galois group is a measure of how complicated the roots of an equation are. For example if your roots are all rational numbers, then the only allowed permutation of the symbols in any equation (with rational coefficients) satisfied by the roots will be the identity permutation, and so the Galois group has just has one element: the identity element (we call this the 'trivial' group). On the other hand, the Galois group of $x^5 - x + 1 = 0$ is *all* of S_5, and we know because there are ways of determining the Galois group without having to find the roots first. Now for Galois' big result: the Galois group contains within it subgroups, and they themselves contain within them subgroups, and so on all the way down to the smallest subgroup (the one with just the identity); we could say there is a certain hierarchy within the Galois group. If those subgroups fit inside one another in just the right way, then the equation is solvable by radicals; if the subgroups of the Galois group do not have the right hierarchy, then it is not. For any polynomial of degree $1, 2, 3$

or 4 the subgroups of the Galois group *always* fit inside each other just right, and so those equations are always solvable. However when we hit degree 5, sometimes the Galois group of a polynomial does have this special hierarchy (like $x^5 + x^4 + 1 = 0$) and sometimes it does not (like $x^5 - x + 1 = 0$); thus some quintics are solvable and some are not.

I would love to say Galois was lauded by the mathematical community, carrying him aloft on their shoulders, chanting his name, but sadly it was not to be. Possibly because of his unusual introduction to Mathematics, Galois had several gaps in his knowledge and failed the entrance exam for the École Polytechnique twice [68]; his first two submissions to the Académie were passed to Cauchy who lost them [119]; his third was given to Fourier who died before he could read it and it was subsequently lost; and tragically Galois' father committed suicide in 1829 over a matter of honour in a disagreement with a local Jesuit. Galois felt his work was being suppressed by the establishment due to his political beliefs, and he drifted more into the republican movement, getting himself arrested several times. It was while in prison Galois heard his third submission to the Académie was rejected by Poisson, who in fairness saw value in Galois' work, he just wanted Galois to present his ideas in a more coherent manner. This seems to have been the last insult for Galois; totally disillusioned, he was challenged to a duel he could not reject (there is much speculation as to the reason of the duel; some say it was over an unrequited love, others that he wanted his death to spark a republican revolt). The night before the duel, convinced he would not survive it, Galois wrote a long letter to his friend and fellow mathematician, Auguste Chevalier, in which he gave a summary of his main results and instructed Chevalier to have them published for all to see; Weyl famously said *this letter, if judged by the novelty and profundity of ideas it contains, is perhaps the most substantial piece of writing in the whole literature of mankind*, high praise indeed! Galois was perhaps a victim of his times: matters of honour, affairs of the heart, duels at dawn. He was shot in the stomach and died the next day from his wounds, aged 20, his body buried in an unmarked grave. What a waste.

The story of Galois, the fiery genius struck down by passion and misfortune, caught the zeitgeist and he has become something of a tragic hero over the years, his legend getting gradually embellished as these things tend to do (from impertinent and arrogant hothead who bit the hand that fed him, to innocent and naïve wunderkind, a *martyr to his genius* [4]). True to his charge, Chevalier had Galois' letter published straight away but it was largely ignored; the letter ended with *I hope there will be some people who will find it to their advantage to decipher all this mess* [114] and no-one rushed to the task. Eventually Liouville realized how profound Galois' work was, and after filling in several gaps published Galois' letter and *mémoire* in 1846. It was perhaps with Jordan's 1870 *Traité des substitutions et des équations algébriques*, giving credit to Galois, that he was finally accepted to the mathematical pantheon.

While Galois had answered the long standing question of solvability of polynomials, perhaps even more important was his development of the notion of a 'group' to do so, and as the 19th century wore on mathematicians realized that groups, actually, *are* all around. Let me give some examples to try and show why we use words like 'structure' and 'universal' when we discuss group theory.

Let's begin with number systems. The set of integers, denoted $\mathbb{Z}$, together with the operation of addition, is a group. We can easily check if the necessary properties are satisfied: when you add two integers you get another integer, the integer '0' plays the role of the identity element (when you add zero to any integer you get the same integer), every integer has an inverse (when you add 3 and -3 you get the identity, 0, so -3 is the inverse of 3 and vice versa). We denote this group $(\mathbb{Z}, +)$, the set of integers with addition as operation. On the other hand, the set of integers with multiplication as operation is *not* a group; yes when you multiply two integers you get another integer, and yes '1' plays the role of the identity, but now there is no inverse: the inverse of 3 for example would need to be $\frac{1}{3}$ (because when you multiply 3 and $\frac{1}{3}$ you get the identity, 1), but $\frac{1}{3}$ is not an integer.

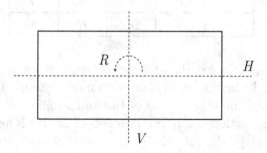

You might have heard that groups are the mathematical way of describing symmetry, so let's see what this means in a geometrical context:

consider the rectangle. If you reflect the rectangle through the horizontal dashed line above, the rectangle before the reflection is identical to the rectangle afterwards; we call this a 'symmetry' of the rectangle. The same is true if we reflect the rectangle through the vertical dashed line, or if we rotate the rectangle 180° about the center point. Remembering that 'doing nothing' is also a symmetry (the rectangle looks the same before and after we do nothing), we see there are 4 symmetries of the rectangle, which we label as

$$E = \text{do nothing,}$$
$$H = \text{reflection through the horizontal line,}$$
$$V = \text{reflection through the vertical line,}$$
$$R = 180° \text{ rotation about the center.}$$

We can compose symmetries, by which we mean do one of these operations and then do another, and the result will again be one of these four symmetries. For example, if we reflect through the horizontal line, and then reflect through the vertical line, this is the same as if we had rotated about the center, or $V \circ H = R$. Rather than go through all the compositions one at a time, Cayley in the 1850s [68] described a table of compositions, now called a 'Cayley table'. We read it like you would a book: left to right, top to bottom. So we take the element that labels a row, and compose it with each element that labels a column:

	E	H	V	R
E	E	H	V	R
H	H	E	R	V
V	V	R	E	H
R	R	V	H	E

We can immediately see that the set of symmetries of the rectangle, together with the operation of composition, is a group. This is because every element of the table is one of the four symbols E, H, V, R which means the composition of any two elements of the set is another element of the set; E plays the role of the identity, and since there is one 'E' in every row this means every element has an inverse. Notice that if we were to draw an imaginary line along the diagonal of this table, we would see that the table of symmetries has a symmetry itself (very meta). This is because, for this group, the order of composition does not matter: $V \circ H = H \circ V$ for example. Groups with this property are called *abelian*, after Abel who we briefly mentioned previously.

Now here's an interesting thing: suppose instead of letting E, H, V, R represent the symmetries of a rectangle, we let these symbols represent the elements of the Galois group of the equation $x^4 - 2x^2 - 3 = 0$, which we listed previously in (7.4). Now when we construct the Cayley table for that group, we would get the *exact same thing* as the table above. More carefully: we can identify every element of one group (group of symmetries of a rectangle) with an element of the other group (Galois group of a quartic polynomial), and vice versa, such that the composition of elements is preserved; in this sense these two groups are *the same*.

Perhaps now you are starting to see why group theory is so central in mathematics: symmetries of a rectangle and roots of polynomials are completely different contexts, but somehow the notion of a group has captured the underlying structural similarities of both settings, giving us a universal language to contemplate the intrinsic properties of things, stripping away the distracting details of context. One more example I think which we will draw on again later in Part IV; the description here is largely inspired by Kuga's *Galois' Dream* [71].

As a young man, I would go for long rambling walks around my home city, Dublin. I would start at the GPO (see Figure 7.1), go walking about, sometimes returning to the GPO and leaving again, before finally returning to where I started. On these long walks, imagine an invisible thread tied to one of the columns of the GPO and then unravelling behind me, like Theseus making his way through the labyrinth. At the end of the day, suppose I want to pull up the thread behind me, gather it all back in, before heading home. The problem is that right in the middle of the city is a large spire, 120 metres tall (the 'stiletto in the ghetto' we liked to call it), and the invisible thread could get snagged on the spire and stop me from gathering it all in, depending on the path I had taken; group theory will help us understand this better.

Suppose I took a map and drew on it the route I walked some day, with arrows denoting the direction I walked in, and labeled the GPO O. Each walk would then be an element of this set:

$C = \{$set of all oriented curves in the plane that start and end at $O\}$.

I could 'compose' walks, as in do one walk that started and ended at the GPO and then do another; this would then be a walk that started and

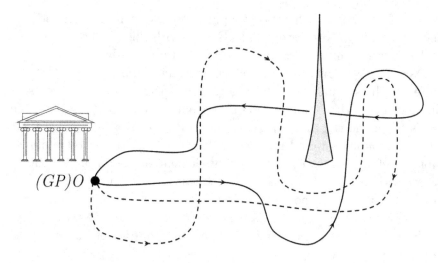

Figure 7.1: Two paths we could walk around Dublin that start and end at the GPO; the arrows indicate the direction in which we walk.

ended at the GPO, or in other words, the set C is closed (composing two elements of C gives another element of C). My main interest is in gathering the thread back in at the end of the day, and as I pull on the thread this means the curve in the plane representing my path will slowly and continuously deform; in fact as long as I can continuously deform one path into another I will consider them 'equivalent'.

Now some paths can be continuously deformed all the way back to the point O; I will call these paths 'null', and if the route I walked some day was a null path that means I will be able to gather up the thread before going home, for example the dashed curve in Figure 7.1 is null. In the set C, null paths play the role of the identity element: if you compose any path with a null path, then this is equivalent to the original path (since you can just shrink away the null part).

What's more, if an element of C is a route I walked in a particular direction, then the inverse of this element is simply the same path walked in the opposite direction (because when you compose those two paths you get a null path). In summary, the set C with the operation of composition is a group, called the *fundamental group*, due to Poincaré (1895 [95]). What sort of group is it?

Let's label all the null paths C_0. The non-null paths are the ones that get snagged on the spire when I try and gather up the thread; when I pull on the thread as much as I can you will see that the thread will be wrapped around the spire a certain number of times. We label the paths that wrap around the spire n times C_n, where n is an integer (let's say n positive corresponds to being wrapped clockwise around the spire, and n negative anti-clockwise, so the undashed path in Figure 7.1 would be labeled C_{-1}). Finally if you compose a path that wraps around the spire n times with a path that wraps around the spire m times, you get a path that wraps around the spire $n + m$ times. So what I am saying is that we can identify the elements of C with the elements of $\mathbb{Z}$, and composing elements of C is the same as adding elements of $\mathbb{Z}$; hence the fundamental group of Dublin is $(\mathbb{Z}, +)$.

It turns out my choice of where to put the point O is irrelevant when it comes to the fundamental group, so really I can take C to be the set of *all* long rambling walks in Dublin that start and end at the same place; even Leopold Bloom's famous meander is an element of C (different spire in those days!).

The spire is on the Northside of Dublin; suppose the Southsiders (they think they're so bleedin' great) decided to put up their own spire, how would this effect the fundamental group of Dublin? Quite a lot actually. Some paths I can shrink back to the point O, which are 'null' as before, and the inverse of any path is still the same route walked in reverse. The tricky thing now is when you pull on the thread as much as you can, it can snag on one spire or the other, or indeed some complicated combination; how do we quantify this? Let's let the letter N denote the thread wrapping once clockwise around the Northside spire, with N^{-1} for anti-clockwise; and similarly S and S^{-1} for clockwise and anti-clockwise around the Southside spire. Now a path might be something like this

$$N^2 S N^{-1} S,$$

which means: the thread wraps twice clockwise around the Northside spire, one clockwise around the Southside spire, then once anti-clockwise around the Northside spire, and finally once more clockwise around the Southside spire (see Figure 7.2). This is called the *word representation* of the path, because we construct it out of letters, and really paths can be as complicated as you like, by just taking some finite combination of N, S, N^{-1} and S^{-1}.

We can do some simplification: in the word $SNN^{-1}S$ the N and N^{-1} will 'cancel', to leave just S^2 (I encourage you to try drawing the picture!), but we can't simplify $SNSN^{-1}$ since we can't rearrange the order of the letters. To compose two walks you just join their words together, and this makes another word which then represents a walk;

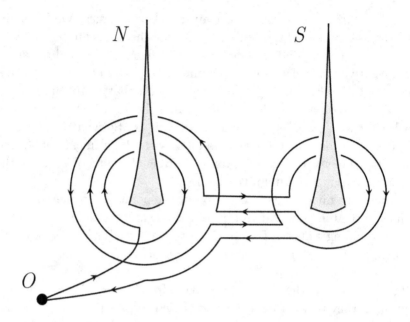

Figure 7.2: The pub crawl from hell; a path with word representation $N^2SN^{-1}S$.

to find the inverse of a word, you write all the letters in reverse order, and replace any N or S with N^{-1} and S^{-1} and vice versa. This set (taking some symbols and constructing words from those symbols and their inverses), together with the operation of composition, is called a 'free group', first introduced by von Dyck in 1882 [68]. The complexity of a free group is measured by the number of symbols used to make the words, called 'generators', so the fundamental group of Dublin with two spires, N and S, is the free group with two generators.

Though the fundamental group has been defined with a set of curves, it isn't really about the curves themselves as such; rather it is about *the space the curves live in*. There is something fundamentally different about Dublin with one spire and Dublin with two, and we know because they have different fundamental groups. But now that we have done the hard work, we get so much more for free: let's think about surfaces like the sphere and the torus. We can fix a point O, and then consider all the curves that lie in that surface that start and end at O. For example, what is the fundamental group of a sphere? Looking at Figure 7.3 we see that on a sphere any curve, no matter how loopy, can be shrunk back

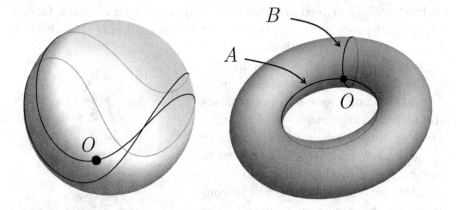

Figure 7.3: The sphere and torus with some curves that start and end at the same point.

to O, and so all curves are null; the fundamental group of the sphere is just the trivial group.

On the other hand, there are curves on the torus which you can only shrink so much before they get 'snagged' and can shrink no more (remember the curves must stay on the surface of the torus). In Figure 7.3 I show two examples, labeled A and B, and it can be shown that *any* non-null curve on the torus is some combination of A loops and B loops. We can therefore represent them with a word, much like Dublin with two spires, only now there is an important difference: on the torus we can rearrange the order of symbols in the word [42], so any word cancels down to something of the form $A^n B^p$ where n and p are integers. Now the composition of two curves, say $A^n B^p$ and $A^m B^q$, is just $A^{n+m} B^{p+q}$; as such we can identify elements of the fundamental group with a *pair* of integers, and composing loops is equivalent to adding those integers separately. Thus the fundamental group of the torus is two copies of $(\mathbb{Z}, +)$, which is often written as $\mathbb{Z}^2$.

Inspired by the success of group theory, other so-called 'algebraic structures' were introduced. For example in our discussion of the roots of polynomials we saw how it was important to choose coefficients which had two operations: addition *and* multiplication (whereas groups only have one), and this lead to the definition of rings and fields, but there are many more: modules, monoids, vector spaces, even 'algebras', all are abstract notions defined in terms of sets, operations and axioms. One of the greatest proponents, and to some a founding figure, of this Abstract Algebra, was Emmy Noether.

It is often said the Noether had the misfortune to be both a woman and a Jew in pre-war Germany, however she also had the fortune to be born at a time when university education was opening up to women and for her father to be an established mathematician, in Erlangen. Her early research was full of computations and equations, which she later criticized as 'crap' and 'a jungle of formulae' [33] (she was famously forthright: 'a loyal friend and severe critic'), and in 1915 moved to Göttingen after the invitation of Hilbert. While the laws of the university prevented women from having a fully paid position (Hilbert is said to have admonished his male colleagues who blocked her promotion by saying "the senate is not a bathhouse!") it was in the 1920s that she established herself as one

Figure 7.4: Emmy Noether, 1882-1935 [61].

of the leading mathematicians of the day, finally gaining the recognition she deserved by being invited to speak at the 1932 International Congress of Mathematicians. Described by Einstein as *the most significant creative mathematical genius thus far produced since the higher education of women began*, she was a poor teacher but instead spent all her waking hours talking mathematics with colleagues and students, famed for making spontaneous, prophetic comments. Aleksandrov [68] says how, on being introduced to the elements of Topology, she immediately observed that *it would be worthwhile to study directly the groups of algebraic complexes and cycles of a given polyhedron and the subgroup of the cycle group consisting of cycles homologous to zero* (I can just imagine the other mathematicians in the room exchanging awkward glances, "why didn't we think of that?"). This insight of Noether's, as expressed in her *Ableitung der Elementarteilertheorie aus der Gruppentheorie* (1926 [110]), lead to one of the great success stories of the 20th century: Algebraic Topology, which we will see in Chapter 15.

In 1933, after the Nazis came to power, Noether had her right to teach withdrawn. This was at least partly due to her socialist leanings as well as her Jewish heritage, although she was completely uninterested in politics and is certainly more deserving of the romantic description of 'child of mathematics' than Galois; while waiting for her appeal against her dismissal she continued hosting mathematical discussions even though some of her students sat before her in their SA uniforms. She emigrated to the US and took up a position at Bryn Mawr College, her first fully paid position [114] (although she still sent half her salary to her nephew back in Germany), but it was short-lived; she died in 1935. In the many eulogies after her death, Alexandrov [33] said *she became the creator of a new direction in algebra and the leading, the most consistent and prominent representative of a certain general mathematical doctrine - all that which is characterized by the words 'abstract mathematics'*, and it is this abstract high-level viewpoint that has provided deep and transformative insights across the world of mathematics: the classification of geometries such as Euclidean and non-Euclidean using group theory, the solvability of differential equations using differential Galois theory, the extension of complex numbers to higher dimensions, the resolution of Fermat's last theorem and much more. Abstract algebra spread beyond mathematics into areas where symmetry is an obvious concern (like crystallography, tilings, even the Rubik's cube) and perhaps most importantly into Physics, not least Noether's own work on the relationship between symmetry and conservation laws in mechanics (see Chapter 11), but also general relativity, field theory and particle physics all benefited from an abstract framework; if you pick up a book on quantum mechanics you will find it unintelligible unless you know some group theory (and even then I am not making any promises). The penetrating gaze of abstract algebra was like the eye of Sauron, stripping away the skin and scars to reveal the bones and organs of how things work, a universal and unifying theory of structure and form. Noether was the purest expression of this totally conceptual and idealized mindset, as van der Waerden said *she was unable to grasp any theorem, any argument unless it had been made abstract and thus made transparent to the eye of her mind* [33]. She had braved the dizzying heights and drank in the view, a far-reaching vista few can fathom, and now as we take our turn and tread the winding stair we might see a flicker of a shadow ahead, hear a distant footstep; a cold hand beckons, climb higher! In time we will see the whole world at once.

CHAPTER 8

Linear Algebra

I T HAS BEEN SAID that the 19th century was the century of Calculus, and the 20th century the century of Topology, but the 21st century will be the century of Linear Algebra; this is because Linear Algebra is the natural framework in which to comprehend large sets of data, a realm which will be increasingly important as we enter the age of Artificial Intelligence and Machine Learning. Linear Algebra has a very practical problem solving side so most engineering and science students will be shown matrices, determinants, inverses and solving linear systems, but under the bonnet Linear Algebra is an abstract and formal system whose principal notion is high-dimensional spaces. As I will describe below there are many instances in history where mathematicians were using ideas and techniques which we now associate with Linear Algebra, however the main window of creation was the 1840s and 50s with Hamilton, Grassmann and Cayley.

On the 16th of October, 1843, William Rowan Hamilton was walking by the Royal canal, not far outside Dublin, with his wife. He writes [119] *although she talked with me now and then, yet an undercurrent of thought was going on in my mind.* Hamilton had, for perhaps the previous decade, been trying to extend complex numbers to three dimensions. He saw the complex number, $a + bi$, as a combination of the units 1 and i, with $i^2 = -1$. Could we not write numbers like $a + bi + cj$ which are now a combination of $1, i$ and j, with $i^2 = -1$ and $j^2 = -1$ but i is not equal to j? Hamilton pictured this in his mind like three perpendicular axes labeled $1, i, j$, and $a + bi + cj$ as an arrow pointing to the point with coordinates (a, b, c), and he called this a *vector*. Little did he know his efforts were doomed to failure - he would come down to breakfast, and

DOI: 10.1201/9781003455592-8

his sons would ask "Well, Papa, can you multiply triplets?" He would reply, with a sad shake of his head, "No, I can only add and subtract them". The problem is: if you do try to multiply two numbers of that form, what should ij be? Should it be 1? Or -1 perhaps? Or ji? Every choice lead to nonsense. But on that day, walking by the canal, he says *an electric circuit seemed to close, and a spark flashed forth!* He realized he needed a *fourth* dimension, numbers of the form $a + bi + cj + dk$, and he took a knife and carved on the stone of a bridge they were passing:

$$i^2 = j^2 = k^2 = ijk = -1,$$

the fundamental formulae for *quaternions* (see Figure 8.1). While previous mathematicians had dabbled with higher dimensions, adding dots to the ends of formulae, this was perhaps the first time that the *necessity* of a higher dimensional space was seen.

Hamilton used the term *scalar* for the coefficient of the 1 unit in a quaternion, and the product of two quaternions has two parts: the scalar part (called the 'scalar product') and the vector part (called the 'vector product'). While both have geometrical meaning, Hamilton had wanted to develop a notion of vectors in arbitrary dimensions, what he termed *polyplets* [68], and sure enough almost immediately after Hamilton's discovery of quaternions his colleague John Graves derived an eight dimensional version of complex numbers, called *octonians*, using similar methods to Hamilton, however that is where the trail runs cold: the

Figure 8.1: A commemorative plaque on Broome bridge, Dublin, where Hamilton famously carved his quaternion formulae [93].

approach only works for dimensions 1,2,4 and 8. Instead a rival view developed as we will see below, where vectors were lists of numbers viewed as coordinates in higher dimensional space, and the scalar and vector product were defined in a combinatorial fashion, with no need to appeal to complex numbers. This approach had the advantage that it could be extended to vectors in any dimension, and eventually quaternions faded into the background, but the circuit had been closed: mathematicians were now emboldened to imagine and work with higher dimensional spaces regardless of any physical motivation.

I don't know about you, but I find it hard to draw pictures in 7-dimensional space; that's ok, we don't need to, the notation and terminology will handle that for us, but nonetheless it is helpful to have a picture in mind so we will often draw diagrams in 2 and 3 dimensions. Suppose we have a point in 3-dimensional space with coordinates (a, b, c), which means we have some point designated the origin, O, and a Cartesian set of axes. We can think of this in three equivalent ways: a set of coordinates (a, b, c), an arrow drawn from the origin to that point,

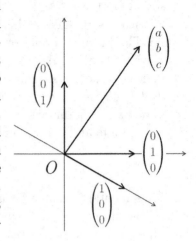

or the list of numbers we typically write vertically as $\begin{pmatrix} a \\ b \\ c \end{pmatrix}$; to all of these we can associate the word 'vector'. It seems reasonable to write

$$\begin{pmatrix} a \\ b \\ c \end{pmatrix} = a \begin{pmatrix} 1 \\ 0 \\ 0 \end{pmatrix} + b \begin{pmatrix} 0 \\ 1 \\ 0 \end{pmatrix} + c \begin{pmatrix} 0 \\ 0 \\ 1 \end{pmatrix},$$

where the three vectors on the right are arrows that point in each of the x, y, z coordinate directions (these are often labeled i, j, k following Hamilton's quaternion notation). Expressions like the one on the right are called 'linear combinations', and the letters a, b, c, called 'scalars', are usually taken to be real numbers (which we denote $\mathbb{R}$), but they don't have to be. The 'linear combination' is one of those everyday notions that has been around forever we just didn't think of calling it that, for example when you have a meal in a restaurant you are presented with a bill which is a linear combination of the prices of items ('you had two of these and three of those...').

In 1844, Hermann Grassmann published *Die lineale Ausdehnungslehre* (the Science of Linear Extension, [68]) where he imagined you take a set of n objects, which we might label $e_1, e_2, \ldots, e_n$, and construct the set of *all linear combinations* of them; for example the set would include $2e_3 - 4e_7$ and $17e_9 + e_{10} + e_{11}$ and so on. He defined how to add elements of this set (as you might expect: adding $e_1 + 2e_2$ and $e_1 - e_2$ gives $2e_1 + e_2$) and how to multiply elements of the set by a scalar (again no surprises: if you multiply $3e_7$ by 4 you get $12e_7$). The crucial thing to note is that the set of all linear combinations is 'closed' with regards to those two operations; you can't 'get out' of the set by adding elements of the set or multiplying by a scalar, since this is just giving you another linear combination. Grassmann called this abstract notion 'the space of' $e_1, \ldots, e_n$, and he went further: he defined how to interpret products like $e_1 e_2 e_4$ in terms of lower products like $e_1 e_2$ and $e_2 e_4$, and those then in terms of e_1 and e_2 and so on. He also defined a scalar product of two elements of the set: you multiply the corresponding coefficients and add them up. Grassmann's work was unfortunately very hard to read and went largely unnoticed by the mathematical community, which is a shame because so many of our modern conventions are already there: if we let e_1, e_2 and e_3 be $\begin{pmatrix} 1 \\ 0 \\ 0 \end{pmatrix}, \begin{pmatrix} 0 \\ 1 \\ 0 \end{pmatrix}$ and $\begin{pmatrix} 0 \\ 0 \\ 1 \end{pmatrix}$, then the set of linear combinations is of course the set of all points with coordinates $\begin{pmatrix} a \\ b \\ c \end{pmatrix}$, but this can now be taken as *the definition* of 3-dimensional space. We call this 'Euclidean space' (to distinguish from the non-Euclidean spaces we saw in Chapter 2), and since the scalars are typically real numbers we denote 3-dimensional Euclidean space with $\mathbb{R}^3$. Now there is a natural extension: let $e_1, e_2, \ldots, e_n$ be the vectors with all zero entries except for a '1' in one slot (e_i having a '1' in the ith slot), then

$$
\mathbb{R}^n = \left\{ \begin{array}{c} n\text{-dimensional} \\ \text{Euclidean space} \end{array} \right\} = \left\{ \begin{array}{c} \text{Set of all linear} \\ \text{combinations of} \end{array} \begin{pmatrix} 1 \\ 0 \\ \vdots \\ 0 \end{pmatrix}, \begin{pmatrix} 0 \\ 1 \\ \vdots \\ 0 \end{pmatrix}, \ldots, \begin{pmatrix} 0 \\ 0 \\ \vdots \\ 1 \end{pmatrix} \right\}
$$

It helps if you try not to think of this as a physical space; let me give an example.

There have been many models proposed to quantify an individual's personality, and one well-respected model is called OCEAN: openness, conscientiousness, extraversion, agreeableness, neuroticism [117]. You take a survey, and you get a score against each of those five criteria; we could write your score as (o, c, e, a, n), and this can now be interpreted as a point in 5-dimensional space, $\mathbb{R}^5$. Suppose two of your friends, Carl and Emmy, also do the survey, they will also be a point in $\mathbb{R}^5$. Who are you more like, Carl or Emmy? Or to put it another way, which point in $\mathbb{R}^5$ is your point closest to? There are many different ways to measure distance in $\mathbb{R}^n$, but the simplest is a generalization of the Pythagoras theorem: the square root of the sum of the squares of the differences in each coordinate.

Now suppose you have a huge survey of the entire population, thousands and thousands of people, each with their own OCEAN score; this can be visualized as a cloud of points in $\mathbb{R}^5$. It might be reasonable to

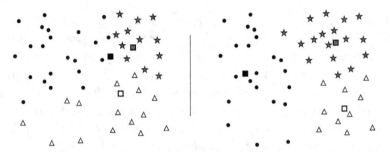

Figure 8.2: On the left the data points are assigned to three different categories due to an initial random choice of centers (squares), and on the right the final clustering using k-means.

think that there are some broad categories of personality type: introverted, psychopathic, creative or whatever. Can we use this large set of data to group people together? This is called 'clustering', and a very common method for doing this is called 'k-means', proposed by Lloyd in 1957 but only published in 1982 [76]. It goes like this: let's say we want to divide our data into three clusters. We randomly choose three points in $\mathbb{R}^5$ and call them the black, gray and white 'centers'. Now we go through all the points in the cloud and for each one measure the distance to each of the centers; we label each point a black dot, gray star or white triangle depending on which center it is closest to (see Figure 8.2). Next we take all the black points and calculate their 'mean': we add them all up, and divide by the number of black points. This will give us a new point in $\mathbb{R}^5$; how do we know? Because this adding up of points and multiplying by a number is a linear combination, and $\mathbb{R}^5$ is the set of *all* linear combinations. We call the mean of the black points the new black center, and we do the same for the gray and white points to get the new gray and white center. Then we repeat the process: go through all the points and measure the distance to the new centers, update the colors of the points appropriately, then get the mean of all the new black/gray/white points and they become the new black/gray/white centers, and iterate several times. Typically things will settle down into three well-defined clusters, and these are our three personality categories. Now if you meet a new person and you want to decide if they are an introvert, psychopath or creative, you get their OCEAN score and see which cluster they fit into (that is, which center they are closest to). k-means is a standard introductory algorithm in Machine Learning [17], and the basic operations this method relies on are simple Linear Algebra.

Just like the notion of a group was extended beyond permutations to instead refer to any set which satisfies certain criteria, in 1888 Giuseppe Peano in his *Calcolo geometrico* [68] made the following definition: we have a set of 'things', with two operations (adding two elements of the set and multiplying an element by a scalar), such that the set is closed under those operations (any linear combination of elements of the set is again an element of the set); there were also some properties that needed to be satisfied, like an identity scalar and an inverse element. Peano called such a construct a 'linear system', but today we call this a *vector space* [5]. Perhaps the most important idea now is that you can always find a set of elements, called a *basis*, such that any other element in the space is a linear combination of that basis; the number of elements in the basis Peano defined as the *dimension* of the vector space. You can see this is somewhat the reverse of Grassmann, who started with a basis and then constructed the set out of it. A basis for $\mathbb{R}^3$ could be $\begin{pmatrix} 1 \\ 0 \\ 0 \end{pmatrix}$, $\begin{pmatrix} 0 \\ 1 \\ 0 \end{pmatrix}$ and $\begin{pmatrix} 0 \\ 0 \\ 1 \end{pmatrix}$, and since the basis has three elements then $\mathbb{R}^3$ is 3-dimensional, but then we knew that already didn't we? The beauty of abstract definitions like Peano's is that it applies to *any* set of things with those properties, and now we have a language to get a handle on those new things. For example, think of the set of all quadratics, i.e. all expressions of the form $ax^2 + bx + c$. If you add two quadratics, you get another quadratic; if you multiply a quadratic by a scalar, you get another quadratic. The set of *all* quadratics is a vector space. How big is this space? Well $1, x$ and x^2 are quadratics, and any quadratic is a linear combination of $1, x$ and x^2, so they form a basis for the space of quadratics; since there are three of them, this vector space is 3-dimensional. Bigger than the plane (2-dimensional), smaller than OCEAN-space (5-dimensional). We can go further - pick two quadratics, how close are they? We can define a notion of distance in the space of quadratics, just like we did in OCEAN-space; we can even say when two quadratics are 'perpendicular', using Grassmann's scalar product, but let's pull it back and give a different example of a context where the notion of basis is useful.

The figure on the right below is a *graph*, nodes connected by edges; in fact it is a *directed graph*, because the edges have arrows indicating direction. Graphs are everywhere: road networks, train lines, pipes, veins and arteries; indeed anything connected, it doesn't need to be physical: webpages, social networks, even mathematical ideas (K_∞!). Graphs might contain 'loops', by which I mean a path that starts at one node, travels along some edges, and returns to the same node. The question is: how many loops does the graph have?

Certainly very many, for example you could start at B, move along the edge a, then along the edge b, and then along the edge c but in the opposite direction to the arrow, and you end up back at B. A natural way to denote this loop would be $a + b - c$, where the minus is because we moved against the arrow on c. Other loops might be $c - d$, $d - c$, $e + f + d$, $b + e + f + a$, and so on. But how do we capture *all* the loops? Let's look at it this way: the edge a starts at node B and ends at node A, so we write it as $A - B$; the edge e starts at C and ends at D so we write it $D - C$; the edge f is $B - D$ and so on. Now take the path $a + b - c$, and replace a with $A - B$, replace b with $C - A$ and replace c with $C - B$. We will get

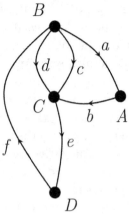

$$(A - B) + (C - A) - (C - B),$$

but this all cancels down to 0; this is how we know $a + b - c$ is a loop. Try it with any loop we saw already: when you replace the edge label with the difference between the labels of the nodes at the ends (terminal minus initial), everything cancels to 0. On the other hand, a path like $a + b + e$ will not cancel down to 0: it is not a loop. So the set of all loops is all the combinations of edges which cancel down to 0 when you replace the edge label with the node labels, or in symbols: it is the set of all expressions like

$$\alpha a + \beta b + \gamma c + \delta d + \epsilon e + \zeta f$$

which cancel to 0 when we make the replacement. So let's do that:

$$\alpha(A - B) + \beta(C - A) + \gamma(C - B) + \delta(C - B) + \epsilon(D - C) + \zeta(B - D).$$

We can expand out the brackets and gather terms together:

$$A(\alpha - \beta) + B(-\alpha - \gamma - \delta + \zeta) + C(\beta + \gamma + \delta - \epsilon) + D(\epsilon - \zeta),$$

and the only way this can all cancel down to 0 is if each of the coefficients is equal to zero, which gives us then 4 equations:

$$\alpha - \beta = 0, \quad -\alpha - \gamma - \delta + \zeta = 0, \quad \beta + \gamma + \delta - \epsilon = 0, \quad \epsilon - \zeta = 0. \quad (8.1)$$

This is called a *linear system* of equations, because each equation is a linear combination of the unknowns. How to solve it?

One of the first things an undergraduate mathematics student sees is the method of Gaussian Elimination [111], a systematic way of solving linear systems. Naming the method after Gauss is a bit generous (his main contribution was devising a special notation for the process), since not only were people like Newton doing these manipulations in the 1600s, but as we saw in Chapter 5 the Chinese *Jiuzhang suanshu* described an algorithm for solving linear systems perhaps 2000 years before Gauss. It goes like this: the labels for the unknowns, α, β and so on, are largely irrelevant, what matters are the coefficients, so we store them in a rectangular table like so:

$$\begin{pmatrix} 1 & -1 & 0 & 0 & 0 & 0 \\ -1 & 0 & -1 & -1 & 0 & 1 \\ 0 & 1 & 1 & 1 & -1 & 0 \\ 0 & 0 & 0 & 0 & 1 & -1 \end{pmatrix}, \tag{8.2}$$

one row for each equation, with the entries in each row being the coefficient of α, β etc. in the corresponding equation (the right hand sides of the equations are all zero so we ignore them here). Now the process is to combine rows so as to introduce more zeroes (we 'eliminate' terms). For example if we replace 'row 2' with 'row 1 plus row 2', we get

$$\begin{pmatrix} 1 & -1 & 0 & 0 & 0 & 0 \\ 0 & -1 & -1 & -1 & 0 & 1 \\ 0 & 1 & 1 & 1 & -1 & 0 \\ 0 & 0 & 0 & 0 & 1 & -1 \end{pmatrix}.$$

Now we could replace 'row 3' with 'row 2 plus row 3', and keep going in this fashion until we reach the following simple form:

$$\begin{pmatrix} 1 & -1 & 0 & 0 & 0 & 0 \\ 0 & -1 & -1 & -1 & 0 & 1 \\ 0 & 0 & 0 & 0 & -1 & 1 \\ 0 & 0 & 0 & 0 & 0 & 0 \end{pmatrix}. \tag{8.3}$$

The linear system represented by this table is equivalent to the original one in (8.1), but it is easier to solve since there are more zeroes. Notice the last row is all zeroes, this is because the original four equations were really just combinations of these three equations.

But we still have a linear system to solve, and there are only three equations for our six unknowns $\alpha, \beta, \gamma, \delta, \epsilon, \zeta$. This means there are not enough equations to pin down a unique solution, there are instead infinitely many (as we perhaps suspected). Look at the third row in (8.3); it says $-\epsilon + \zeta$ is equal to zero, so ϵ is equal to ζ. If we let ζ be t, then this equation implies ϵ is also t. Now the second row in (8.3) says $-\beta - \gamma - \delta + \zeta = 0$; if we let γ and δ be s and r respectively, then this equation means $\beta = t - s - r$. Finally the first row says $\alpha = \beta$, so $\alpha = t - s - r$ also. Thus our full solution can be written as follows:

$$
\begin{pmatrix} \alpha \\ \beta \\ \gamma \\ \delta \\ \epsilon \\ \zeta \end{pmatrix} = \begin{pmatrix} t - s - r \\ t - s - r \\ s \\ r \\ t \\ t \end{pmatrix} = t \begin{pmatrix} 1 \\ 1 \\ 0 \\ 0 \\ 1 \\ 1 \end{pmatrix} + s \begin{pmatrix} -1 \\ -1 \\ 1 \\ 0 \\ 0 \\ 0 \end{pmatrix} + r \begin{pmatrix} -1 \\ -1 \\ 0 \\ 1 \\ 0 \\ 0 \end{pmatrix}.
$$

Now look at the three vectors on the right; each one identifies a specific loop in the graph: the first vector (the one multiplied by t) is the loop $a + b + e + f$, the second vector is the loop $-a - b + c$ (or perhaps more easy to see it in the graph if we write it as $c - b - a$) and the third vector is the loop $d - b - a$ (see below). We could have spotted those loops ourselves, but this is saying something stronger: *all* the loops in the graph are linear combinations of these three simple loops, because *all* the solutions of (8.1) are linear combinations of these three vectors. As such the three simple loops identified form a basis for the space of all loops, and therefore the space of loops on this graph is 3-dimensional.

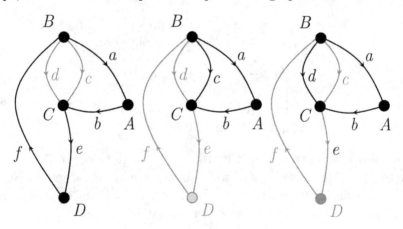

Maxwell, motivated by electrical circuits, called this the 'cyclomatic number' [49], and using the Euler characteristic we will meet in Part IV of this book there is a quick formula:

$$\left(\begin{array}{c} \text{Dimension of space of} \\ \text{loops in a graph} \end{array} \right) = \Big(\text{ \# edges } \Big) - \Big(\text{ \# nodes } \Big) + 1.$$

There is another distinguishing feature of Linear Algebra which we have hinted at in the last section: the matrix. While most undergraduate students of science and engineering will see a little matrix theory, determinants and inverses and so on, primarily as formulae to help solve linear systems, matrices were first studied (as objects in their own right) via the notion of a 'linear transformation'. We are familiar with functions like x^2, which means: give me a number, like 3, and I will return another number, 9. A function has an input and an output, but they need not be numbers; they could be, for example, vectors. So you might have a function where the input is a vector in $\mathbb{R}^3$, and the output is a vector in $\mathbb{R}^2$, like in the image below.

This is too big a problem to think about, so in 1858, in his *Memoir on the theory of matrices* [68], Arthur Cayley studied the simplest kind of function we can think of: *linear* functions, but in this context they tend to be called 'linear transforma-

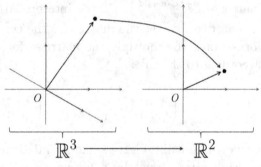

tions'. The simplest case is when the input is a point in the plane, $\mathbb{R}^2$, with coordinates x, y, and the output is another point in the plane, let's say the coordinates are X, Y. To say the function is 'linear' means we can write the new coordinates as linear combinations of the old coordinates:

$$X = ax + by, \quad Y = cx + dy,$$

where a, b, c, d are (typically real) numbers. Doing lots of this type of computation gets cumbersome after a while, so Cayley represented this linear transformation with a table of the coefficients:

$$\begin{pmatrix} a & b \\ c & d \end{pmatrix}.$$

He called this a *matrix*, the name coined by his long-standing friend and colleague James Sylvester. Cayley and Sylvester, mild and mercurial respectively, both had varied careers: though they studied at Cambridge, Sylvester could not graduate because he was Jewish, and instead graduated from Trinity College Dublin (which reserved its religious bigotry for Catholics). They both went into the law, and even though Cayley was a barrister for 14 years he published over 300 mathematical papers in this time. They worked so extensively on the theory of 'invariants' (combinations of the elements of matrices which do not change under certain transformations) they were known as 'the invariant twins'; this study of invariants was the same field in which Noether created her 'jungle of formulae'.

   Suppose there is a second linear transformation, whose matrix is now $\left(\begin{smallmatrix} \alpha & \beta \\ \gamma & \delta \end{smallmatrix}\right)$, Cayley asked: what happens when you compose these two linear transformations? Do one, then do the other? Now the point (x, y) goes to $(ax + by, cx + dy)$, but then *that* point goes to $(\alpha(ax + by) + \beta(cx + dy), \gamma(ax + by) + \delta(cx + dy))$. This means that the composition of two linear transformations is another linear transformation, with coefficient matrix $\left(\begin{smallmatrix} \alpha a + \beta c & \alpha b + \beta d \\ \gamma a + \delta c & \gamma b + \delta d \end{smallmatrix}\right)$. Cayley interpreted this as a multiplication rule for matrices: to get the matrix for the composition of two linear transformations, you multiply the matrices for those linear transformations, according to the rule:

$$\begin{pmatrix} \alpha & \beta \\ \gamma & \delta \end{pmatrix} \times \begin{pmatrix} a & b \\ c & d \end{pmatrix} = \begin{pmatrix} \alpha a + \beta c & \alpha b + \beta d \\ \gamma a + \delta c & \gamma b + \delta d \end{pmatrix}.$$

Cayley used capital letters like A and B to denote matrices, and matrix multiplication ('rows by columns') is now something students all over the world learn, complicated by the fact noticed by Cayley that the order in which you compose linear transformations (and hence multiply matrices) matters: AB is not the same as BA.

   The matrices we just saw had two 'rows' (the horizontal lists of numbers) and two 'columns' (the vertical), and that is because we were thinking about a linear transformation from $\mathbb{R}^2$ to $\mathbb{R}^2$. If our linear transformation was from $\mathbb{R}^n$ to $\mathbb{R}^m$, then the matrix representing it would have m rows and n columns, which we write as $m \times n$ (so in the sketch on the previous page the matrix would be 2×3). The matrix multiplication rule above can be easily extended to matrices of any size (with some caveats), and in particular Cayley used matrices to write a linear system of equations in a more compact form; for example

$$
\begin{aligned}
3x + 2y + z &= 39 \\
2x + 3y + z &= 34 \\
x + 2y + 3z &= 26
\end{aligned}
\qquad \text{can be written} \qquad
\begin{pmatrix} 3 & 2 & 1 \\ 2 & 3 & 1 \\ 1 & 2 & 3 \end{pmatrix}
\begin{pmatrix} x \\ y \\ z \end{pmatrix} =
\begin{pmatrix} 39 \\ 34 \\ 26 \end{pmatrix}.
$$

This particular linear system is over 2000 years old: it is Problem 1 Chapter 8 of the *Jiuzhang suanshu* (see (5.1)). As for notation, to avoid confusion with coordinates vectors are often written $\underline{x}$, $\vec{x}$ or $\boldsymbol{x}$ (I prefer the bold) so the system above can be written

$$ A\boldsymbol{x} = \boldsymbol{y} $$

where $A = \begin{pmatrix} 3 & 2 & 1 \\ 2 & 3 & 1 \\ 1 & 2 & 3 \end{pmatrix}$ is the matrix of coefficients, $\boldsymbol{x} = \begin{pmatrix} x \\ y \\ z \end{pmatrix}$ is the 'unknowns', and $\boldsymbol{y} = \begin{pmatrix} 39 \\ 34 \\ 26 \end{pmatrix}$. Cayley hoped to work with matrices just like we do with numbers to construct and solve matrix equations (as well as multiply he also defined how to add matrices and multiply by a scalar), but he quickly found a snag: if you wanted to solve the equation $ax = y$ you could multiply both sides by $\frac{1}{a}$, which is the 'inverse' of a; to solve the matrix equation $A\boldsymbol{x} = \boldsymbol{y}$ we would need to multiply across by the 'inverse' of A, but unfortunately not all matrices have an inverse.

Nonetheless matrices have a rich structure, for example: if you add two $m \times n$ matrices, you get another $m \times n$ matrix, and when you multiply an $m \times n$ matrix by a scalar you get another $m \times n$ matrix; in other words, the set of all $m \times n$ matrices is a vector space (whose dimension is $m \times n$). But also look at the rows of a matrix: they are all lists of numbers, and so they are all vectors; the set of all linear combinations of the rows of a matrix is a vector space, called the 'row space'. The same goes for the columns of a matrix: they define a vector space called the 'column space'. Perhaps surprisingly, the row space and column space of a matrix always have the same dimension, even though they live in larger spaces of different dimensions: if you look at the matrix in (8.2), the rows live in $\mathbb{R}^6$ and the columns in $\mathbb{R}^4$, but the row space and column space both have dimension 3.

There's more: taking all the $n \times n$ matrices that have an inverse, they form a group, called the 'general linear' group, and this group has lots of subgroups which have been studied more deeply than you can imagine. Also, depending on the structure and context, there are all sorts of types of matrices, a veritable zoo of matrix species: symmetric and skew-symmetric, stochastic and orthogonal, tridiagonal and diagonalizable; matrices can be nilpotent, singular, self-adjoint, and involutory, I could go on all day! Alas, Statement A. Instead, let me give you a 21st century example of an application of Linear Algebra.

If I were to show you a picture of an animal and asked you to identify it as a cat or a dog, you would get it right pretty much all the time; humans are really good at this sort of thing. Trying to teach a machine to do it on the other hand is much harder, so why are humans so good at it? Do we have some criteria for 'cat-ness' or 'dog-ness' that we check against? No, really what has happened is that over the course of your life you have seen lots of animals that you were reliably informed is a cat, and lots of animals you were told is a dog, so now when you see a new animal you are comparing it to the previously classified animals; you have 'trained' your brain to classify cats and dogs. I will describe here a machine learning technique that uses Linear Algebra to do the same: use data to train a machine to classify images.

By 'image' we mean a grid of pixels, where each pixel has a color, but to keep things simple we will suppose that the grid of pixels is 5×5, and the images are black and white; as such any image can be written as a list of 25 (grayscale) numbers, or to put it another way: an image is a vector in $\mathbb{R}^{25}$, let's call it $\boldsymbol{x}$. We want to feed into the machine a list of 25 numbers and for it to spit out a classification: cat or dog. A nice way to visualize this is with a graph [17]: on the left are the 'input' nodes, one for each element of my image vector $\boldsymbol{x}$ so 25 in all, and on the right is the 'output' node, just one because we want our algorithm to spit out 'cat' or 'dog'. Let's say the output, y, will be a number scaled to between -1 and $+1$, so a negative output we call a 'cat' and a positive output a 'dog'. What we are setting up here is a function, with an input $\boldsymbol{x}$ and an output y, and the simplest function we can think of is a *linear* function, so let's suppose the output is a linear combination of the input vector elements

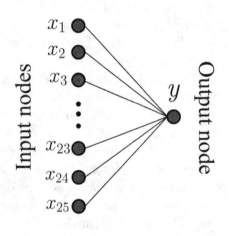

$$y = a_1 x_1 + a_2 x_2 + \ldots + a_{25} x_{25}. \tag{8.4}$$

Here the coefficients of the linear function, the a_i, are called 'weights' as they determine how much each x_i contributes to the decision. The big question is: what value should the weights take?

Let's train our model: we take lots and lots of images which we definitely know are dogs, convert each image into a list of 25 numbers (x_1, x_2 and so on), sub them into (8.4) and set $y = +1$ each time; we do the same with lots and lots of images of cats, only now we set $y = -1$. This will give us a linear system of equations where the unknowns are the weights a_i, and we go ahead and solve using the marvellous methods of Linear Algebra. Now we have values for the a_i, when we come across a *new* image we just convert it into a list of 25 numbers x_i and sub them into (8.4); if the output is positive we say 'dog', if negative we say 'cat'.

The problem with all this is that we will have a pretty rubbish classifier. Just like people designed airplanes by looking at how birds fly, if we are to design an artificial intelligence we should look at how brains think. Like a hugely complex graph, the brain has neurons (the nodes of the graph) connected by synapses (the edges), and we are going to take two lessons from the brain: firstly, experiments of Hubel and Weisel [59] in the 60s showed that neuronal networks are arranged in hierarchical layers, one layer feeding into the next.

To try and keep the numbers modest, I will suppose there are two intermediate layers between the input and output layer, and that they also have 25 nodes (although in practice there could be more layers and many more nodes). As before we suppose that the values at the nodes in each layer are lin-

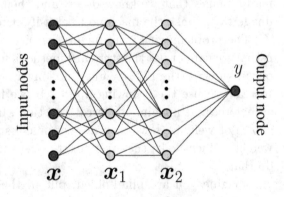

ear combinations of the nodes at the previous layer: if x_1 is the 25 values of the nodes in the first intermediate layer, then $x_1 = Cx$, where C is the 25×25 matrix of those linear combinations, and similarly $x_2 = Bx_1$ where B is the matrix of linear combinations relating the nodes at the second intermediate layer to the first; finally $y = Ax_2$, so altogether

$$y = Ax_2 = ABx_1 = ABCx.$$

The weights are the entries in the matrices A, B, C only now A has 25 entries but B and C both have 625 entries, so altogether there are 1275 weights to find, and this huge increase in the number of parameters to play with is the benefit of inserting those intermediate layers (although in practice many of those parameters will be zero).

The second lesson from the brain is how neurons work: they don't just pass on every signal that comes their way, instead they wait until a certain input threshold is reached before they 'fire'. The standard way [17] to incorporate this into our approach is, rather than having one layer's nodes being linear functions of the previous, to instead use 'piece-wise' linear

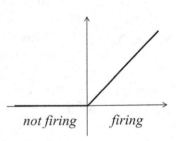

functions (in effect: if the value at a node is negative, we suppose the node has not 'fired' so we set it to zero). We have arrived at what is known as a 'convolutional neural network', and this combination of layers and firing make CNN's very flexible, powerful and efficient. Training a machine to identify cats from dogs may not seem like a huge contribution to civilization; however we could instead feed the algorithm many many MRI images which we know contain cancerous tumors, and many many images that we know don't, and then when we have a *new* MRI image we can ask the machine to classify it for us, perhaps saving lives.

The problem with introducing the piece-wise linear 'firing' function is that we cannot use the standard methods for solving linear systems, as our unknowns (the 1275 weights) no longer solve a strictly linear system. Instead we use the following approach: we feed in the training images as before, and imagine an 'error function', defined by: for a particular choice of weights, how accurate is our classifier? Then we look for the weights which make the error function as small as possible. How do we do that?

Imagine you are blindfolded, put in the back of a helicopter, and dropped off on the side of a mountain somewhere; you need to walk down into the valley to safety. Suppose you can only see the ground by your feet, but no further. One strategy would be turn yourself all the way round and decide which direction is 'downhill', then start walking in that direction, every now and again stopping to check which direction is downhill. This is essentially how we minimize the error function, only now we need to walk down a mountain in 1275 dimensional space! This is where Linear Algebra meets Calculus.

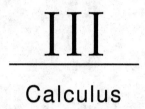

Calculus

CHAPTER 9

The Beginnings

P UT SIMPLY: Calculus is the greatest idea Mathematics ever had. Given how central calculus has become in science and engineering in general it is hard to overstate how important it is, but I don't think anyone would argue with me if I said that without calculus there would be no footprints on the Moon, without calculus there would be no artificial intelligence or nuclear power plants, and without calculus we wouldn't have smart phones to entertain ourselves with funny cat videos all day. The path to calculus was a convoluted one, to paraphrase Grabiner [50]: *first calculus was used, then discovered, explored and developed, and only then, defined.* This twisting path is much documented (I recommend Strogatz [113], Nahin [86] and Hairer and Wanner [53]) and I will try to sketch the big ideas as they arose using as little jargon and technicality as possible; while we will see some of the great successes in the application of calculus to the world around us, to some degree the story of calculus is that of a long hard-fought battle with the bogeyman of mathematics and philosophy alike: infinity.

The classical Greeks were wary of infinity, like superstitious sailors who would not dare venture out into the empty reaches of the mathematical map (*there be monsters*, only these monsters were giant sea serpents that ate their own tails). Their compass was a subtle distinction, due to Aristotle [114], between *potential* and *actual* infinity; if we look back to Chapter 4 where we described Pappus filling the space between two circles with smaller and smaller circles, he considered an unending process where we can always add more circles (potential infinity), rather than the final state where we have added infinitely many circles (actual infinity). To Archimedes (3rd century BCE Syracuse), this meant care

DOI: 10.1201/9781003455592-9

needed to be taken in presenting his results to posterity, and there were many results to present; some would consider him to be the greatest mathematician of antiquity (if not all time), and more of his writings survive than any other great mathematician of the time.

A good example would be his proof in *Measurement of a circle* [41] that the area of a circle is equal to the area of a right-angled triangle that has the radius as one short side and the circumference as the other. He said: suppose the area of the circle is *greater than* that of the triangle; we label this second area K. We must therefore be able to construct a polygon inscribed in the circle whose area is less than that of the circle, but more than K; note the polygon may need to have lots of sides, but still only a finite number. Archimedes then showed easily that such a polygon would have an area less than K, so the original assumption (the area of the circle is greater than the triangle) must be false. He then supposed the area of the circle is *less than* that of the the triangle, and with a circumscribed polygon showed that that must also be false; the only conclusion then is that the area of the circle is *equal* to that of the triangle. This 'method of exhaustion' (originally due to Eudoxus [109]) was perfectly rigorous, and only required potential infinity to work; the problem was you needed to already know the result you were trying to prove (in this case, you needed to already know what the area of the circle was). Where did the original idea come from?

Archimedes spoke of a mysterious 'method', which he didn't want to share widely and as such was lost to the world when the Greek heyday passed; or so it was thought. Archimedes had in fact confided his method in a letter to his colleague Eratosthenes in Alexandria, which was then copied by scholars along with several other classical texts in a manuscript. In the middle ages, probably due to a suspicion of Greek writing and arcane drawings, the manuscript was written over then forgotten about in a Palestinian monastery for centuries before being rediscovered at the start of the 20th century, the original Greek (see above [25]) being barely legible underneath. Modern imaging techniques (coupled with Linear Algebraic methods) make the background stand out, and we can now read the words of Archimedes, hidden for two thousand years, as he reveals his Method.

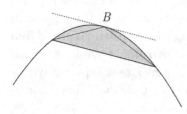

Consider the parabola cut by a line to make what is called a 'parabolic segment', the shaded region on the left. Archimedes set out to measure the area of this segment, but since everything was proportion we need something to refer it to, so we let the line drift across in a parallel fashion until it touches the parabola at a point, B. This point B and the base of the segment form a triangle, and Archimedes wanted to compare the areas of the parabolic segment and this triangle.

Here we go: draw a vertical line at A, and extend it until it meets the line that is tangent to the parabola at C. Extend the line from C to B through K and on to H, so K is the midpoint of CH. Being mechanically minded (bit of an understatement there) Archimedes imagines the line CH as a balance bar with K as pivot,

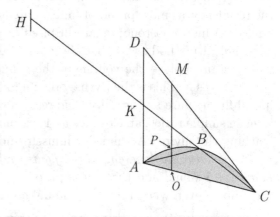

and takes a vertical line segment of the parabolic segment, like OP, and moves it to the point H, where he argues that this line segment is then balanced by the corresponding line segment of the larger triangle OM, quoting results from Euclid's *Elements* and his own previous work. He then says that *since the [parabolic] segment is made up of all the straight lines like OP within the curve*, the whole parabolic segment, cut up into infinitely many infinitely thin lines and shifted to H, is balanced by the larger triangle ADC (whose center of gravity is known), and therefore the area of the parabolic segment is one third that of ADC, whose area is in turn four times the area of the triangle ABC; therefore the area of the parabolic segment is $\frac{4}{3}$ times the area of the triangle ABC.

Today we can admire the power of creativity in this construction, but Archimedes feared the obvious criticism: he had invoked *actual* infinity. How can we compare infinitely many of one thing against infinitely many of another? And what even can a line segment weigh, being infinitely thin? He was quick to point out to Eratosthenes that he didn't consider this approach a 'proof', *but that [the] argument has given a sort of indication that the conclusion is true*; in other words, this approach suggests what the area of the segment is, and he can then go ahead and prove it is so using the standard 'method of exhaustion' technique (which he did also). But to the modern eye we can appreciate it is precisely this attack on actual infinity that is at the heart of Calculus: divide something up into infinitely many pieces, say something about those pieces, then add them all up again.

Some would say Archimedes anticipated integral calculus by 1800 years or so, but that is a bit of an overstatement in my opinion. Nonetheless Archimedes was clearly a master of his art: the result he was most proud of (and wanted engraved on his tombstone), again inspired by his 'mechanical method', was that the volume of a sphere is precisely $\frac{2}{3}$ the volume of the cylinder that encloses it. This is the Pythagorean mindset in full flight: pure and idealized figures, proportions in simple ratios, an elegance and symmetry  reaching the divine. Archimedes himself said [41] *these properties were all along naturally inherent in the figures referred to, but remained unknown...* I would add that these properties will always be inherent in the figures, even when they become unknown again.

I hate to burst the bubble, but the fact is simple ratio is not enough to capture proportion even for the most basic of constructions, and what could be more basic than the ratio of a circle's perimeter to its diameter. Classical Greeks didn't pay this ratio much attention; Euclid doesn't mention it at all, and there was no special name or symbol, our use of π being a recent convention (Jones 1706, π for 'periphery' [53]). Archimedes attempted an 'exhaustion' type approach to find this ratio, sandwiching the perimeter of a circle between inscribed and circumscribed polygons with more and more sides; going as far as 96-sided polygons he found π is bound in the range

$$\frac{223}{71} < \pi < \frac{22}{7},$$

whose midpoint is accurate to 3 decimal places. The
difficulty of the calculation put many of his successors off, although around 1600 Ludolph van Cuelen
calculated the first 35 decimal places of π by using
polygons with 6×2^{60} sides! But this Herculean effort
was from a bygone age, and while van Cuelen was
toiling away, Viète was laying down the new method:
convert a (geometrical) problem to an algebraic one

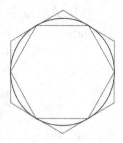

by using symbols for the unknowns, then solve. At the end of his text
he wrote in capital letters *nullum non problema solvere*: giving solution
to any problem. It was time to put this claim to the test.

Mathematicians of the 1600s had inherited two overarching geometrical problems from the Greeks: finding the area of a region bounded
by curves, and finding the tangent lines to curves (we saw both of
these in Archimedes' work above). While the Greeks had studied curves
like circles and conic sections, the joining of Geometry and Algebra by
Descartes and Fermat meant that any curve could be captured by an
equation, but also any equation represented a curve, so now there was
an overwhelming myriad of curves to study, even those described by simple equations like $y = x^2$, $y = x^3$, $y = x^4$ and so on. For example, what
is the area under the curve $y = x^n$, between $x = 0$ and some other point,
say $x = B$? While initially some like Cavalieri [68] tried to answer this
question by extending classical techniques (you could use Archimedes'
result above to find the area under $y = x^2$, since Descartes tells us this
is a parabola), it was Fermat (1636 [53]) who attacked this problem in
a truly modern way.

Let θ be a number slightly
smaller than 1. If we repeatedly
multiply B by θ we get a sequence
of points on the x-axis, getting
closer and closer together as we
approach the origin. At each of
these points we raise a rectangle
that meets the curve $y = x^n$ (in
the diagram we suppose $y = x^2$).
Look at the tallest rectangle on
the right: its width is $B - \theta B$,
and its height is B^n, therefore its
area is $B^{n+1}(1 - \theta)$. Now look at
the rectangle beside it: its width

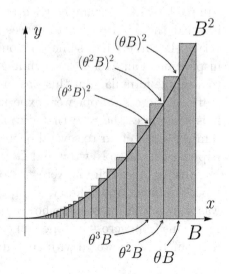

is $\theta B - \theta^2 B$, and its height is $(\theta B)^n$, so its area is $B^{n+1}(1-\theta)\theta^{n+1}$. The next rectangle has area $(\theta^2 B - \theta^3 B) \times (\theta^2 B)^n$ which can be written $B^{n+1}(1-\theta)\theta^{2n+2}$ and so on, which means the sum of the areas of all the rectangles is

$$B^{n+1}(1-\theta)\left(1 + \theta^{n+1} + \theta^{2n+2} + \dots\right).$$

The infinite sum in the round brackets is a geometrical series, and already in the 1300s Oresme [16] knew how to sum these: the sum of the areas of the rectangles is

$$B^{n+1}\frac{1-\theta}{1-\theta^{n+1}}.$$

The denominator has a well known factorization: we can write this expression as

$$B^{n+1}\frac{1-\theta}{(1-\theta)(1+\theta+\theta^2+\dots+\theta^n)}.$$

Now Fermat said: we want the rectangles to get narrower and narrow so they better approximate the area under the curve, which means we want θ to get closer and closer to 1. So he did something which made everyone shift uncomfortably in their seat: he first said that θ is *not* equal to one, so we can cancel the $1-\theta$ in the numerator and denominator, but then he said that θ *is* equal to one, so the area under the curve as

$$\frac{B^{n+1}}{n+1}.$$

Fermat was uneasy about this, as were those who followed him (as we will see later), and while he tried to justify this step there was no denying that the main reason to proceed in this way was that it *worked* so very well: rather than some tortuous bespoke construction needing the inspiration of an Archimedes, finding the areas under curves was reduced to a simple formula. But this was just the beginning, and lots of mathematicians across Europe were experimenting with constructions like this; it is sometimes the case that during an exciting time of invention the same idea can occur to several people at once, and this is precisely what happened to Isaac Newton and Gottfried Leibniz.

Newton and Leibniz were the classic odd couple: Newton was famously difficult and solitary, aloof, guarded in sharing his work; Leibniz was a gregarious diplomat who travelled widely, meeting leading mathematicians from across Europe and exchanging ideas. Newton was lauded in his lifetime, showered with honours and buried in Westminster Abbey;

Leibniz died penniless, only his secretary attending his funeral. But in other ways they were very alike: scholars of science and philosophy as well as mathematics, they both invented languages based on scientific principles [87], and they both claimed to have invented Calculus. Here is a typical example of the approach they took:

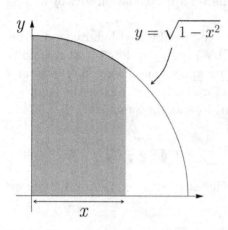

Figure 9.1: Using power series to calculate π.

Consider a circle of unit radius, centered on the origin. From Archimedes we know the area of such a circle is π, and from Descartes we know the equation of the circle is $x^2 + y^2 = 1$; this means the area under the curve $y = \sqrt{1 - x^2}$ between 0 and 1 is $\pi/4$, but let's think about the area under the curve up to the point x. Newton was a master of the *power series* (which we met in Chapter 6, like a polynomial but with infinitely many terms). While there was an Indian school which developed power series, perhaps as early as the late 1300s by Madhava [68], Newton was the first European to (re)discover power series by experimenting with the binomial theorem but he did not publish his results, which meant Mercator (1668) was the first in print (this situation will be a recurring theme later). Newton and Leibniz worked with power series like we would with numbers, adding and multiplying, squaring and inverting. The series for $\sqrt{1 - x^2}$ is

$$1 - \frac{x^2}{2} - \frac{x^4}{8} - \frac{x^6}{16} - \frac{5x^8}{128} - \cdots$$

which means we can say

$$\left(\begin{array}{c} \text{Area under the} \\ \text{curve } y = \sqrt{1 - x^2} \end{array} \right) = \left(\begin{array}{c} \text{Area under} \\ \text{the curve} \end{array} \quad y = 1 - \frac{x^2}{2} - \frac{x^4}{8} - \frac{x^6}{16} - \cdots \right),$$

but we know from Fermat how to find the area under the curves given by $y = x^n$, so we can write

$$\left(\begin{array}{c} \text{Area under the} \\ \text{curve } y = \sqrt{1 - x^2} \end{array} \right) = x - \frac{x^3}{6} - \frac{x^5}{40} - \frac{x^7}{112} - \frac{5x^9}{1152} + \cdots$$

Finally letting x run up to 1 we have

$$\left(\begin{array}{c}\text{Area of a}\\\text{quarter circle}\end{array}\right) = \frac{\pi}{4} = 1 - \frac{1}{6} - \frac{1}{40} - \frac{1}{112} - \frac{5}{1152} - \cdots$$

from which we can get a better and better approximation to π by adding more and more terms.

I wouldn't recommend it though, you would need hundreds of terms to get to Archimedes' accuracy; instead there are lots of other infinite series we can derive for π using similar approaches. Newton (1660s) and Leibniz (1670s) derived power series for $\arcsin(x)$, which is the length of the heavy arc at the top of the shaded region in Figure 9.1:

$$\arcsin(x) = x + \frac{x^3}{2.3} + \frac{3x^5}{2.4.5} + \frac{3.5x^7}{2.4.6.7} + \cdots$$

which they then inverted to get the infinite series expression for $\sin(x)$:

$$\sin(x) = x - \frac{x^3}{3!} + \frac{x^5}{5!} - \frac{x^7}{7!} + \frac{x^9}{9!} - \cdots \tag{9.1}$$

We can see in the plot below how taking more and more terms in the series expansion gives a better and better approximation to $\sin(x)$:

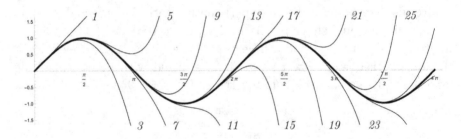

In 1674 Leibniz discovered the series for $\arctan(x)$ (more properly called the Madhava-Gregory-Leibniz series for its discoverers in c1400, 1671 and 1674 respectively), from which in 1719 de Lagney used this expression

$$\pi = 2\sqrt{3}\left(1 - \frac{1}{3.3} + \frac{1}{5.3^2} - \frac{1}{7.3^3} + \frac{1}{9.3^4} - \cdots\right)$$

to calculate π accurate to 127 decimal places. As I write, we know π to 100 trillion (!) decimal places, and by the time you read this it will probably be even more. To put this in perspective, if we wanted to calculate the circumference of the known universe down to the accuracy of the width of an atom, then 40 decimal places would be plenty. Some people would ask: why bother? These are the same people who would stand on the beach with their arms folded while some poor sod emerged from the surf, steaming and exhausted from having swum the Channel, and say 'why bother?' Luckily for us, people like Newton and Leibniz *did* bother, and now we have the modern age! But I am going off on a tangent. . .

We have seen how the ancients constructed tangents to special curves like circles and ellipses, but with the advent of Analytic Geometry the question was now how to find the tangent to an *arbitrary* curve, without the elaborate bespoke constructions; what was needed was a general method that could be used on curves like $y = x^n$, or even the curve defined by $x^3 + y^3 - 3xy = 0$. It seems fitting that this problem was first addressed by the inventors of Analytic Geometry itself: Fermat and Descartes.

Actually Fermat first described a purely geometrical method *before* he invented Analytic Geometry, but then recast it in algebraic terms, perhaps as early as 1629. This is how he proposed to find the tangent to a parabola: suppose the parabola has equation $y = x^2$, and the equation of the tangent line is $y = mx + c$, where m (slope of the line) and c (y-intercept) are to be found. Fermat first said that if the parabola and line intersect then there are points that satisfy both these equations simultaneously, so solving the equation for the line as $x = \frac{y-c}{m}$ and subbing this into the equation for the parabola we get $y = (\frac{y-c}{m})^2$. This is a quadratic equation, with roots

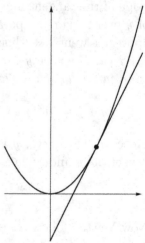

$$\frac{2c + m^2 \pm m\sqrt{m^2 + 4c}}{2}.$$

Now Fermat reasoned that if the line is *tangent* to the parabola then this quadratic should only have one (double) root, which means that what is under the square root sign should be zero; this means $c = -\frac{m^2}{4}$ and after a little algebra we soon get the slope of the tangent $m = 2x$.

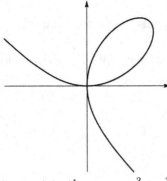

Descartes, as we saw in Chapter 3, described in *La géométrie* [31] the osculating circle, whose radius would be normal to the curve and then the line perpendicular to that would be the tangent line; the problem with Descartes' device was that the resulting algebra was much more complicated than Fermat's. For example he could make no progress on finding the tangent to the curve $x^3 + y^3 - 3xy = 0$ (shown above), so he instead set it for Fermat as a challenge in 1638; Fermat solved it easily.

Newton's approach was much more mechanical or dynamical: he viewed a curve as the path of a particle moving in the plane, and so as time goes by the x and y coordinates are changing; he called them *fluents*, and their rate of change *fluxions* which he denotes $\dot{x}$ and $\dot{y}$. The slope of the tangent line is then the ratio of $\dot{y}$ over $\dot{x}$. For the parabola $y = x^2$ his approach went like this: replace x with $x + \dot{x}o$ and replace y with $y + \dot{y}o$ where o is *an infinitely small quantity* to give

$$y + \dot{y}o = x^2 + 2x\dot{x}o + \dot{x}^2 o^2.$$

Since, on the curve, y is equal to x^2 we cancel the y and x^2 leaving

$$\dot{y}o = 2x\dot{x}o + \dot{x}^2 o^2.$$

Now Newton followed Fermat's lead [114]: he writes *the remaining terms being divided by o there will remain*

$$\dot{y} = 2x\dot{x} + \dot{x}^2 o,$$

but whereas o is supposed to be infinitely little, the terms that are multiplied by it will be nothing in respect to the rest; I therefore reject them and we are left with $\dot{y} = 2x\dot{x}$ which means $\frac{\dot{y}}{\dot{x}} = 2x$. In other words, o is not zero, and then it is.

Though Newton developed these methods in the 1660s, he didn't publish straight away; he wrote some tracts which are known by the shorthands *De Analysi* and *De Methodis* [68], and though he showed them to colleagues they were only actually published decades later, in some cases years after Newton was dead. In fact he himself said *pray let none of my mathematical papers be printed without my special licence* [53]. He would come to regret this: Leibniz published first.

If nature is like a giant book open before us, and the language of that book is mathematics, then notation is the font of that book, and if you don't think font is important then try reaDL*ₙg* this ₛ𝐄ɒ𝘵e𝗡𝞼ɔ̃ᷓ. Leibniz was a master builder of notation, and he focused his attention on the *characteristic triangle* (on the right but see also Figure 3.6). He used the letter d for *differentialis* (Latin for 'difference'), so for example dx is the difference in the x-coordinates of the two points shown. Since the rectangle has width dx and height y, and therefore area $y\,dx$, Leibniz denoted the sum of the areas of lots of these rectangles with an 's', which looked like 'f' at the time

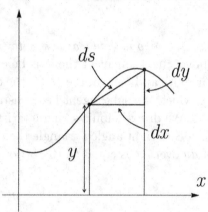

(i.e. the area is f $y\,dx$) and this symbol was gradually elongated to

$$\int y\,dx$$

which we now call 'the integral' of y (see Figure 9.2). Also, in the characteristic triangle we see the slope of the tangent line will be close to

$$\frac{dy}{dx}$$

which Leibniz called the 'differential quotient'. His *Nova methodus* in 1684 [74], arguably the first Calculus publication, had all the rules of his d operator; for example he wrote the product rule as $d(xy) = x\,dy + y\,dx$. He saw dy and dx as small but not zero, so they could be freely treated like fractions; for example in his notation things like $\frac{dy}{dx} = \frac{dy}{dt} \cdot \frac{dt}{dx}$ are obvious. What's more, he introduced the terms *calculus differentialis* and *calculus integralis*, so now we can talk about 'differentiation' and 'integration' rather than 'slope of the tangent' and 'area under the curve'. To the modern reader, it is the work of Leibniz which will feel most familiar (but it could have been very different: he originally went with *omn.y* instead of $\int y$, with $\frac{x}{d}$ instead of dx, and with *calculus summatorius* [16]).

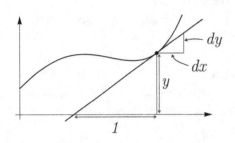

It was becoming clear that these new methods could be used to tackle problems which previously had seemed intractable. In his 1684 paper Leibniz considered, as he put it, *the most difficult and most beautiful [problem] of applied mathematics, which without our differential calculus or something similar no one could attack with any such ease.* Look at the diagram above: the segment of the axis below the tangent line is called the 'sub-tangent', and back in the 1630s the question was posed: what is the curve for which the subtangent has constant length? Let's suppose this length is 1. We draw Leibniz's characteristic triangle, and since it is similar to the other right angled triangle then it immediately follows that the ratio of dy over dx is equal to the ratio of y over 1, in other words,

$$\frac{dy}{dx} = y.$$

This is an equation involving differentials, and so we call it a *differential equation.* This is a first order differential equation, and if you have taken a course in first order differential equations you may have met the methods known as 'separation of variables' and 'the integrating factor'; Leibniz invented both of these in the 1690s.

He went further: differences of differ-ences, for example $d(dx)$, which he de-noted d^2x. Using the diagram on the right he found that the second difference in y was proportional to the square of the dif-ference in x, specifically $d^2y = -y(dx)^2$, which can be written as

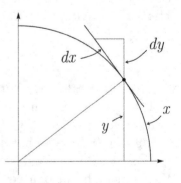

$$\frac{d^2y}{dx^2} = -y, \qquad (9.2)$$

a second-order differential equation (this explains the funny positioning of the 2's in the numerator and denomi-nator). Leibniz supposed the solution was a power series, and when he subbed it in and married up the coefficients he rediscovered the series for $y = \sin(x)$ as we gave it in (9.1).

By now the power of the new calculus was undeniable, and though the original motivation in the development of calculus was geometrical, tangent lines and areas, as we move into the 18th century attention was turned toward the application of calculus to physical problems, and that is where differential equations took center stage; in fact the next three chapters are largely about differential equations and their very many applications.

Almost immediately the logical foundations of Calculus were criticized: how can something be not zero one moment, and zero the next? Leibniz said we could ignore $(dx)^2$ when compared to dx, but how could something be infinitely smaller than something that is already infinitely small? In integration we see the same issue Archimedes tried to hide: if the rectangles have zero width and there are infinitely many of them, how can we calculate their area? The most famous criticism was from Bishop Berkeley in his *The Analyst* of 1734 (whose subtitle begins *A discourse addressed to an infidel mathematician...*). He writes [41] *for when it is said, let the increments vanish, i.e. let the increments be nothing... the former supposition that the increments were something, or that there were increments, is destroyed, and yet a consequence of that supposition... is retained... a false reasoning.* The bishop's point was that it was a bit rich for the newly enlightened scientists to mock the faithful, when they themselves seem to be accepting principles which they could not explain; perhaps they are simply swapping one faith for another? He asks *whether such mathematicians as cry out against mysteries have ever examined their own principles?*

While some responded to these criticisms with dismissal (Johann Bernoulli [53] passes it off as *small beer; who could refrain from laughing at his ridiculous hair-splitting about our calculus, as if he were blind to its advantages*), others, such as Maclaurin in 1742, tried to justify the new approach in *the manner of the ancients* (i.e. the method of exhaustion), but this was contrary to the whole point of Calculus. Lagrange in 1797 tried to do away with any mention of infinitesimals: he said every function (another term due to Leibniz) has a power series, and the coefficients in the series were themselves functions. He called the original function $f(x)$ the *primitive*, and the coefficients *derived functions*, which he denoted $f'(x), f''(x)$ and so on, from which we get our term *the derivative*. The problem with this formulation is it became clear that not all functions which we would like to include in our analysis can be expressed as power series.

The resolution came in the way these things often do: a vague and intuitive notion, gradually refined and made more precise as time went by. The key concept is that of a 'limit'. Already Newton used the word in the sense of a value that some variable approaches closer and closer (indeed this notion was implicit in Archimedes' measurement of the circle), and numerous others (such as d'Alembert and Bolzano) made this idea more precise, but perhaps the definitive contribution was from Cauchy. In 1821 he published his *Cours d'Analyse de l'École Royale Polytechnique* where he gave the modern definitions of limits and continuity, followed in 1823 with *Résumé des Leçons* with the modern definition of the derivative. We will try and describe the ideas intuitively, and the formal statements we will leave for the textbooks.

Suppose we want to find the *limit* of the function x^2 as x gets closer and closer to 2, by which we mean the value that x^2 approaches as x approaches 2 (or x 'goes to' 2); of course we expect this value to be 4. Cauchy introduced the 'lim' notation so we would now write this as

$$\lim_{x \to 2} x^2 = 4$$

by which he meant: we can make x^2 as close as we like to 4 by simply letting x be sufficiently close to 2. Cauchy used the letters ε and δ to describe those two measures of 'closeness', and ε-δ has been melting the brains of mathematics students (mine included) ever since. The key thing to notice is we do not need to let x be actually *equal* to 2, merely be as close to 2 as we want. Note also that the value of the limit, 4, is the same as the value of the function at that point; we therefore say the function x^2 is *continuous* at $x = 2$, it doesn't have a 'jump' there.

Consider now the function $\frac{\sin(x)}{x}$. This function is not defined when $x = 0$, because $\frac{0}{0}$ can take any value you wish; however the *limit* of this function as x goes to 0 *does* exist. Let's use the power series for $\sin(x)$ we have seen already (in (9.1)):

$$\lim_{x \to 0} \frac{\sin(x)}{x} = \lim_{x \to 0} \frac{x - \frac{1}{3!}x^3 + \frac{1}{5!}x^5 + \dots}{x}$$

and remember that we are only taking the limit as x goes to 0, x is not actually equal to 0, so we can divide the top and bottom by x:

$$\lim_{x \to 0} \frac{\sin(x)}{x} = \lim_{x \to 0} 1 - \frac{1}{3!}x^2 + \frac{1}{5!}x^4 + \dots$$

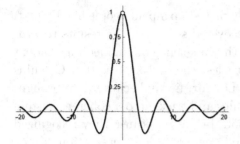

Now we can make this infinite sum as close to 1 as we like, by just taking x to be small enough; as such we have found that

$$\lim_{x \to 0} \frac{\sin(x)}{x} = 1.$$

You can see in the plot on the left that the graph of the function approaches 1 as x gets close to 0 (the empty circle means the function is not defined there).

Now we can finally define the derivative of a function, as you would typically find it in a decent modern text like [6]: if $f(x)$ is a function of x, then the *derivative* of f with respect to x, denoted $f'(x)$, is given by

$$f'(x) = \lim_{h \to 0} \frac{f(x+h) - f(x)}{h} \tag{9.3}$$

(remember we are not just 'subbing in' $h = 0$, we are taking the limit as h goes to 0). This can either be interpreted as a function (like Lagrange did), an instantaneous 'rate of change' (like Newton did) or as the slope of a tangent line (as Fermat and Leibniz did): in the diagram, the slope of the line joining the points with coordinate x and $x + h$ will get closer and closer to that of the tangent line at x as the two points get closer and closer together, i.e. as h goes to 0.

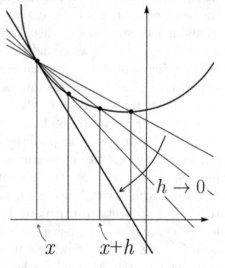

Although it was the best part of 200 years since Fermat intuitively introduced the ideas of calculus, mathematicians had finally put the infinity bogeyman back in the box. In attempting to make calculus watertight, Cauchy (and others of course) had introduced 'rigor' into mathematics; in fact Cauchy's publications in the 1820s could be seen as the beginning of a whole new branch of mathematics which we now call 'Analysis'.

So who invented Calculus? Lagrange said *one may regard Fermat as the first inventor of the new calculus,* and Laplace agreed. By 1660, before Newton or Leibniz had barely picked up a pen, people like Fermat, Barrow, Wallis and Hudde had discovered several of the results regarding differentiation and integration that a science or engineering student might meet in a first course. Nonetheless the consensus is that Calculus should be credited to Newton and Leibniz, firstly because they brought together a disparate collection of rules and tricks into a coherent system equipped with a specialized notation, but also because they recognized the inverse relationship between differentiation and integration, which in Leibniz's notation is almost self-evident:

$$\int \frac{dy}{dx}\, dx = y \; (+ \text{ a constant}),$$

the Fundamental Theorem of Calculus. Newton and Leibniz corresponded, a little, and cordially (although in his letters Newton made sure to say things like *I myself fell upon this theory several years ago*), but when in 1695 Newton heard that in the Netherlands Leibniz is considered the inventor of Calculus, there began a bitter argument between Newton's and Leibniz's followers, not just over who developed Calculus first chronologically but whether one had in fact plagiarised the other. In 1711 the Royal Society set up a commission to resolve the matter, but the president of the Society at the time was one Sir Isaac Newton so we should not be surprised at their findings. 300 years on we can look back and say that both Newton and Leibniz independently 'invented' Calculus (bearing in mind the contributions of those who went before, Fermat in particular), and while Newton made his discoveries before Leibniz (10 years perhaps) it was Leibniz who published first, in 1684 (more bluntly: Newton was scooped in '68 by Mercator and he was scooped in '84 by Leibniz; he didn't want to publish but still wanted the credit). The real outcome of the dispute was a sort of mathematical Brexit where English mathematicians no longer collaborated with their European colleagues, and stuck firmly with Newton's notation (in 1785 the Cambridge tripos still asked students to calculate fluents from fluxions [68]), leading to a 'lost century' for English mathematics. In Europe however Leibniz's notation, superior to Newton's in several respects, was expanded and developed by the Bernoullis, who in turn taught it to the Marquis de l'Hôpital (who published the first Calculus textbook *Analyse des infiniment petits*, 1696), and then Euler himself. Huge leaps forward happened in 18th- and 19th-century France, Switzerland, and Germany, including attacking probably the biggest question of them all.

MENSIS OCTOBRIS A.MDCLXXXIV. 467

NOVA METHODUS PRO MAXIMIS

*& minimis, itemque tangentibus, quæ nec fractas, nec irrati-
onales quantitates moratur, & singulare pro illis
calculi genus, per G. G. L.*

S It axis AX, & curvæ plures; ut VV, WW, YY, ZZ, quarum ordi- TAB.XII.
natæ, ad axem normales, VX, WX, YX, ZX, quæ vocentur respective, v, vv, y, z; & ipsa AX abscissa ab axe, vocetur x. Tangentes sint
VB, WC, YD, ZE axi occurrentes respective in punctis B, C, D, E.
Jam recta aliqua pro arbitrio assumta vocetur dx, & recta quæ sit ad
dx, ut v (vel vv, vel y, vel z) est ad VB (vel WC, vel YD, vel ZE) vocetur d v (vel d vv, vel dy vel dz) sive differentia ipsarum v (vel ipsarum vv, aut y, aut z) His positis calculi regulæ erunt tales:

Sit a quantitas data constans, erit da æqualis o, & d ax erit æqu.
a dx: si sit y æqu. v (seu ordinata quævis curvæ YY, æqualis cuivis or-
dinatæ respondenti curvæ VV) erit dy æqu. dv . Jam *Additio & Sub-
tractio:* si sit z--y + vv +x æqu. v, erit d z-- y + vv +x seu d v, æqu.
d z -- d y + d vv + d x. *Multiplicatio,* d x v æqu. x d v + v d x, seu posito
y æqu. x v, fiet d y æqu. x d v + v d x. In arbitrio enim est vel formulam,
ut x v: vel compendio pro ea literam, ut y, adhibere. Notandum & x
& d x eodem modo in hoc calculo tractari, ut y & dy, vel aliam literam
indeterminatam cum sua differentiali. Notandum etiam non dari
semper regressum a differentiali Æquatione, nisi cum quadam cautio-
ne, de quo alibi. Porro *Divisio,* d^v vel (posito z æqu. ^{v}) d z æqu.

$$+ v \, dy + y \, dv$$
$$y \qquad\qquad y$$
$$yy$$

Figure 9.2: The cover page of Leibniz's *Nova Methodus* of 1684 [74],
the first Calculus publication [61] (full title in English being [114] 'A
new method for maxima and minima, as well as tangents, which is not
obstructed by fractional and irrational quantities, and a curious type of
calculus for it'). On this page we can see Leibniz introducing the rules of
differentiation for *Multiplicatio* and *Divisio*, which we would now call the
'product' and 'quotient' rules. Note also the first use of the '*d*' notation,
as well as the italic 's' which became our integral sign, ∫.

CHAPTER 10

The Solar System

T HE FIRST 'big' questions humans ever asked themselves were probably things like 'where do we go when we die?', 'where do babies come from?' and: what are the stars? And the Sun and the Moon? And why is the day sometimes long and sometimes short? All cultures would have developed their own cosmology, and a common framework we see in several disparate civilizations (for example a Mayan conception [133] on the right) has three layers: the middle layer where we live, an underworld perhaps where the dead live, and an upper realm where the gods live. The middle layer was flat, and the upper layer with the Sun, Moon and stars was domed. This dome would often rotate and hence carry the stars with it. While we are mak-

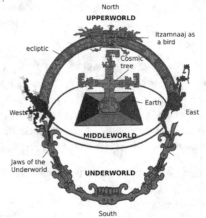

ing sweeping generalizations, we could also say that the first scientists of the ancient world were astronomers whose observations were mostly based around the calendar for agricultural purposes, but they would have inevitably noticed that some stars seem to move in a manner distinct from the other 'fixed' stars. Today we call them 'planets', from the Greek word for wanderer, and it is the Greeks who first developed sophisticated geometrical models of the cosmos; indeed Plato challenged his fellow philosophers to capture the motions of the heavens using *circular motions, uniform and perfectly regular* [68] and two rival models developed. The first was due to Eudoxus (of 'method of exhaustion' fame)

 DOI: 10.1201/9781003455592-10

in the early 4th century BCE: the round static earth at the center of a series of concentric spheres that rotate at a steady rate about their axes, but each axis is attached to a point on another outer sphere which is itself rotating (see the left below). If a planet is attached to even two such spheres this combination of motions can produce very complicated orbits (for example on the right below), and for some planets Eudoxus had four such spheres. Eudoxus didn't suggest that these spheres were physically there, this was just a convenient device with which to make predictions, but in time people came to think that the heavens were foliated by actual physical spheres.

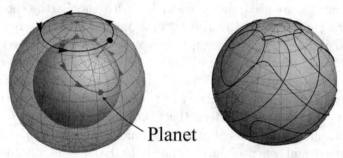

Planet

A rival view was due to Apollonius (of conic sections fame) who instead said that each planet moved on a circle whose center was itself moving on another circle (known as an 'epicycle') whose center was displaced somewhat from the round Earth's position at the center of the cosmos; see the left below. Again all sorts of complex paths can be constructed out of combinations of circles [1], two examples are on the right below. In particular, for the heavy line, the planet (when viewed from the Earth, E) will appear to be moving anti-clockwise at the point A, but clockwise at the point B, which replicated an observed phenomenon most pronounced in Mars (indeed this irregular orbit and bloody color are the reasons the red planet is often associated with war and chaos).

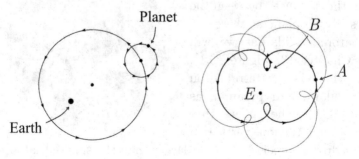

In the 2nd century CE Ptolemy wrote a huge astronomical treatise in 13 books which, like Euclid's *Elements* and Apollonius's *Conics*, replaced all that had gone before it. With a complete description of the epicycle model of the heavens along with detailed calculations of the parameters involved, *the Almagest* was the final word in astronomy for the next 1200 or so years; for example the astronomers of the Islamic golden age used (and extended) the Ptolemaic model. Eventually it started to show its age: discrepancies which were minor in Ptolemy's day accumulated over the centuries and so as time went by doubts started to grow that the model was flawed. In the early 1500s Nicolaus Copernicus became convinced the problem lay with the Earth-centered (or 'geocentric') element, and read how some ancient astronomers [68] had suggested it was in fact the Sun that was at the center of everything; he put forward this 'heliocentric' view in his *De Revolutionibus* of 1543, the first credible astronomical model that did not place the Earth at the center of the universe, contrary to the bible. This didn't cause much controversy at the time, indeed the Copernican model was taught in some Catholic universities, but by the mid-1600s the political winds had changed and when Galileo claimed the Earth was not the center of everything he was accused of heresy, forced to recant, and put under house arrest for the rest of his life. In terms of mathematical development, the Copernican model was no more novel or simple than the Ptolemaic (essentially still planets on epicycles with displaced centers) but putting the Sun at the center provided the inspiration for Johannes Kepler.

Kepler was something of a transitional figure; on the one hand, his methods are those of an evidence-based scientist, on the other his ideas and reasoning were closer to numerology and mysticism. For example, in his *Mysterium Cosmographicum* of 1596 he puts forward the following inspired hypothesis to explain the distances between the spheres of the planets: there are five Platonic solids (as we will see in Chapter 13, Figure 13.1), one of which is the tetrahedron (a triangular based pyramid, as on the right). The sphere that passes through the four vertices of 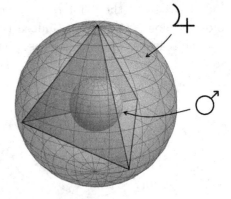 a tetrahedron is precisely 3 times larger than the sphere that sits inside it touching the sides, and the orbit of Jupiter is three times that of Mars, so Kepler said that the sphere of Jupiter contained a tetrahedron which contained the sphere of Mars, which itself contained a dodecahedron which in turn contained the sphere of the Earth, and so on. Why is this the case? Because *God is always a geometer* he said, and we can see in Kepler a true descendent of the Pythagorean 'everything is number' mindset.

However we also see the scientist in Kepler: proud as he was of this cosmic harmony, the model simply did not make predictions that agreed with the data. He spent long years laboring over his calculations, and finally, painfully, Astronomy put aside the Platonic dictat of uniform circular motion: in 1609, Kepler published his *Astronomia Nova* which contained his first two 'laws', the third law following in 1619. Kepler knew that the planets moved more slowly the further they were from the Sun, and this is expressed as 'the line joining the Sun to the planet sweeps out equal areas in equal time': on the left below we see the orbits of two planets over the same time interval. Planet A is closer to the Sun than planet B, so for the two sectors to have the same area the arc AA' must be longer than the arc BB', but if planet A goes from A to A' in the same time that B goes from B to B', it must travel faster. This is now known as Kepler's second law, the first being 'the orbit of a planet is an ellipse with the Sun at a focus'; on the right we see an ellipse with one focus marked and the semi-axes a and b fixing the size and proportion of the ellipse (as we saw in Chapter 1).

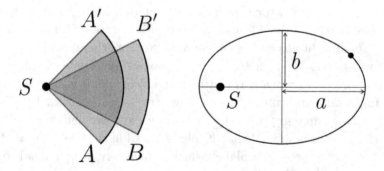

For his third law he says *the ratio which exists between the periodic times of any two planets is precisely the ratio of the $\frac{3}{2}$th power of the mean distances [of each planet from the Sun]*, and this law, like all his laws, is justified empirically: he makes these hypotheses and then checks them against the observations, and sees they are in excellent agreement (within the accuracy of the time). It is ironic that for someone as enamoured with the Classical concept of symmetry and simplicity he was the one to bury the ideal of uniform circular motion (although ellipses and ratios like $\frac{3}{2}$ would have pleased Pythagoras; it was Poincaré's results to come that would have horrified him). But there was still something missing: *why* are the orbits elliptical? What compelled the planets to move in this way?

What did you do during lockdown? In 1665 the plague returned to England yet again, and when Cambridge university closed its doors the 22 year old Isaac Newton returned to the 'family' home. He recalled [109] *In the beginning of the year 1665 I found the Method of approximating series & the Rule for reducing any dignity of any Binomial into such a series. The same year in May I found the method of Tangents of Gregory & Slusius, & in November had the direct method of fluxions & the next year in January had the Theory of Colours & in May following I had entrance into ye inverse method of fluxions. And in the same year I began to think of gravity extending to ye orb of the Moon & . . . from Keplers rule of the periodical times of the Planets . . . I deduced that the forces wch keep the Planets in the Orbs must [be] reciprocally as the squares of their distance from the centres . . . All this was in the two plague years of 1665-1666. For in those days I was in the prime of my age for invention & minded Mathematicks & Philosophy more then at any time since.* Beats making banana bread.

We have seen Newton's reticence about publishing, and it was 20 years before the fruits of his *anni mirabiles* were published as the *Philosophiae naturalis principia mathematica* (mathematical principles of natural philosophy), in 1687. I try to avoid hyperbole in this book, but there are times when it is justified: the *Principia* is easily one of the most important and influential scientific texts ever written. I will give two reasons for making this claim, one direct and one indirect.

Firstly Newton showed that Kepler's laws follow if we suppose that the sun exerts a force on a planet which is inversely proportional to the square of the distance between them (actually he proved Kepler's first law the other way round: if the orbit is an ellipse, then there is an inverse square law [107]). He did much more than that: beginning with three universal laws of motion which hold for *any* bodies subject to *any* forces, he considers gravitational forces (or 'centripetal' forces as he says), several complex problems in conic geometry, circles rolling on circles, oscillations in the shapes of cycloids, bodies passing through fluids and the compression of fluids, the tides and more besides; taken together it is a *tour de force* of the scientific mind turned to the mechanics of nature, and, as we have seen in so many cases already, this text replaced all that went before it such as the work of Galileo and Descartes. While we might recognize in the *Principia* some of the ideas behind Calculus, his proofs of Kepler's laws are entirely geometrical, classical even (it was Johann Bernoulli in 1710 who gave an analytical treatment [107]). Let's see how Newton proved Kepler's second law (equal areas in equal times) but first we need a 'lemma', which is like a mini-theorem: given a triangle in the plane we can slide a vertex or side along parallel lines without changing the area; for example all the triangles below have the same area.

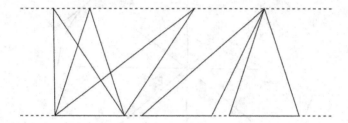

Book I Section II Proposition I Theorem I [62] begins with *suppose the time to be divided into equal parts*; in one unit of time the planet moves from A to B, with the Sun at S (see Figure 10.1). If there were no force acting on the planet, then in another unit of time it will have moved forward by the same amount in a straight line, to the point B' (this is Newton's first law). By our lemma above the triangles SAB and SBB' have the same area.

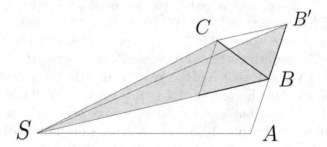

Figure 10.1: Newton's proof of Kepler's second law in the *Principia* [62].

Now suppose instead that at the point B there is a *great impulse* of force on the planet directed toward the Sun; the combination of this force and the planet's otherwise forward motion (heavy lines in Figure 10.1) means it will instead traverse the diagonal of a parallelogram in one unit of time to the point C (Newton's parallelogram law for adding forces, Corollary I after his three laws, is an early example of the vector concept that we have seen several times already in this book). But now the shaded triangles SBB' and SBC have the same area also by our lemma above. Therefore the planet has swept out the triangles SAB and SBC, with the same area, in equal units of time.

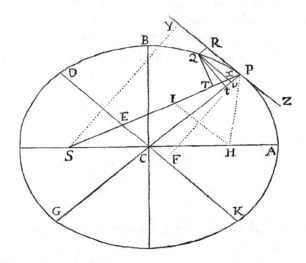

Figure 10.2: An excerpt from Newton's *Principia* [62] for his study of Kepler's first law.

You'll notice this proof makes no mention of the inverse square magnitude of the force, and indeed Kepler's second law holds for *any* centripetal (what we now call 'central') force. The inverse square nature of gravitational force is needed in Newton's proof of Kepler's first law (the orbit is an ellipse with the Sun at a focus), Book I Section III Proposition XI Problem VI, and what Newton gives us is really an elaborate conic section theorem in the vein of Apollonius (in fact Apollonius wrote 8 books in his *Conics*, but only 7 survived; maybe Newton's theorem was in the lost book?). Figure 10.2 shows Newton's diagram, with the Sun at S and the planet at P. Note the conjugate diameters DK and PG, as we saw them in Figure 1.3, and Newton uses the lemma about the areas of parallelograms formed by conjugate diameters which he says has been *demonstrated by the writers on conic sections*: standing on the shoulders of giants. Newton had brought mathematical credibility to Kepler's inspired hypotheses, and given a scientific reason for *why* the heavens appear as they do: as the result of forces between the Sun and the planets (of course we could now ask: but why are there forces? We will come back to that).

A second reason the *Principia* was so transformative is that it established a *method*. If you want to study some mechanical system (let's say particles acted on by forces), you: write down the positions and hence the accelerations of the particles, times by their mass, and set this equal to the forces acting on them. This is Newton's second law, and while he did not phrase it this way in the *Principia* almost everyone sees this law written as

$$F = ma.$$

Let me show you an example: suppose there is a mass m that can move horizontally and the only force acting on the mass is that due to a spring.

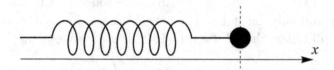

When the mass is at rest it sits at a certain point (the dashed line), and we let x measure the displacement of the mass from that point. In Newton's terminology, x is a *fluent* (it varies with time t), and the velocity is the *fluxion* of this fluent, denoted $\dot{x}$. Moreover the acceleration is the fluxion of the velocity, denoted $\ddot{x}$, so the 'mass by acceleration' part of Newton's second law is $m\ddot{x}$. For the force due to the spring, we use 'Hooke's law' (Hooke was a contemporary of Newton, and in fact they argued over priority of the inverse square law): the force due to the spring is proportional to the displacement from the rest position, and directed toward it. As such we can write the force as $-kx$ where k is the 'spring constant' and takes into account the physical properties of the spring (such as how many coils it has and so on). Now Newton's second law equates these two expressions:

$$m\ddot{x} = -kx. \tag{10.1}$$

The key point to appreciate here is that Newton's second law has lead to a *differential equation* and this is why differential equations have been so central over the last centuries. The k and the m are a bit distracting here, so let's suppose they have a value of 1 for now. If we write this differential equation in Leibniz's notation

$$\frac{d^2 x}{dt^2} = -x$$

then we have seen it before (this is equation (9.2)), and we said how Leibniz used power series to show a solution is $\sin(t)$, but we could also approach it in the following way: we are looking for some function such that when we differentiate it twice we get *minus* that function. $\sin(t)$ certainly has that property, but so does $\cos(t)$ (when you differentiate $\cos(t)$ twice you get $-\cos(t)$). Do we need to keep looking? Are there more solutions? Yes and no: there are in fact infinitely many solutions, but they are all linear combinations of $\sin(t)$ and $\cos(t)$, just like any vector in the plane is a linear combination of $\binom{1}{0}$ and $\binom{0}{1}$, indeed, *exactly* like that, as any book on differential equations will tell you ([103] is a joy, see [66] also). By the 1730s Euler knew how to solve general differential equations of this class [68], by guessing an exponential with an unknown exponent and subbing it in. If we put back the k and the m, Euler's method will tell us the solution to (10.1) is a linear combination of

$$\sin\left(\sqrt{\tfrac{k}{m}}\, t\right) \quad \text{and} \quad \cos\left(\sqrt{\tfrac{k}{m}}\, t\right)$$

and these two trig functions oscillate like a wave as time goes by; in particular the 'frequency' of their oscillation is the combination $\sqrt{\tfrac{k}{m}}$.

This is where the Maths and the Physics really sing together: a physical problem lead to a mathematical equation, and the solution of the mathematical equation tells us how the physical problem will evolve; in this case if you pull the mass past its rest position and let it go then it will oscillate with the 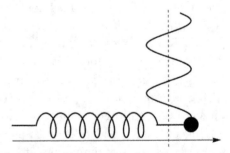 frequency given. This is the classic 'modeling' cycle: you have some real world problem you want to study, you transform it into an idealized mathematical model and use Newton's laws (or similar) to write down some equations, the solutions of which tell you something about the real world problem you are interested in. This 'mathematization of nature' has been phenomenally successful, and Newton's *Principia* was a driving force behind the burgeoning Age of Enlightenment and a foundation stone in our modern secular world; let's not forget that in the time of Newton women were still being burned at the stake as witches.

The problem with all of the models of the solar system so far is they have been 'static', by which I mean the underlying structures are staying the same; a planet may be moving on epicylces but they are the same epicycles for all time. In the 18th century astronomers observed that the orbit of Jupiter was getting smaller whereas the orbit of Saturn was getting larger [91]; how to explain this? Did the radii of the epicycles contract over time, like old rope? Did Jupiter and Saturn repel one another, like a pair of magnets? Perhaps the universe as a whole was like a giant clockwork mechanism, slowly winding down, and needed the occasional attention of an invisible hand to reset it.

In fact Newton had already given us the clue: his second law has an acceleration term which gives a dynamical/time-dependent aspect to the model, plus he presents a 'universal' law of gravitation that *all* elements of matter in the universe exert a gravitational force on all other elements of matter (*Principia* Book III Rule III). Combined, we can write down the differential equations relating forces and accelerations, just like we did for the spring, but now for all the planets in the solar system (and more). Solve those equations, and you know everything.

What we are describing here is an 'n-body' problem, where n is the number of planets, moons, comets and so on you are including. Newton solved the '2-body' problem (also called the Kepler problem), where we suppose the only objects in the universe are the Sun and a planet; the resulting elliptical path of the planet is described by 4 numbers called 'orbital elements' [85] as shown in the diagram: a (semi-major axis) and e (eccentricity) describe the size and shape of the orbit, ϖ (pronounced 'curly pi') rotates the orbit about some reference direction denoted Υ, and v (the 'true anomaly') fixes the actual position of the planet on the orbit (to make the following description clearer we will suppose everything happens in the plane; in 3-d we would need to include two more elements). In the 2-body problem, a, e and ϖ are constants, and v changes with time; also

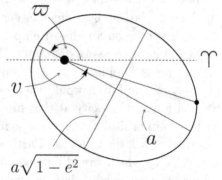

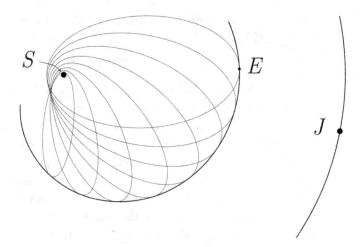

Figure 10.3: A 3-body problem with the Earth orbiting the Sun, perturbed by Jupiter. In thin lines are some osculating ellipses (exaggerated to bring out the dynamics) as the Earth moves clockwise, and we can see how the semi-axis is decreasing while the eccentricity is increasing.

if you know the coordinates (x and y) and velocities ($\dot{x}$ and $\dot{y}$) of the planet at any moment, you can calculate the orbital elements, and vice versa (note the matching of dimension here: $(x, y, \dot{x}, \dot{y})$ and (a, e, ϖ, v) both have 4 numbers).

Now suppose there is a third body, how do we incorporate this? Newton had tried to include a third body in the *Principia* but made little progress: his geometrical approach was doomed to failure because there are no 'standard' orbits, as we will see; Calculus is essential. Let's suppose the three bodies are the Sun, Earth and Jupiter (see Figure 10.3); since the Sun is so much larger than the other two we would expect the Earth to be on an almost elliptical orbit that is 'perturbed' (altered slightly) by the presence of Jupiter. We can approach it this way: imagine everything is happily floating around in space when Jupiter suddenly disappears; the Earth would then be in the familiar 2-body set-up and its position and velocity at that moment would define an elliptical orbit with certain orbital elements. Now imagine instead if Jupiter had in fact vanished at a later time, the Earth would then have a different position and velocity which would define an elliptical orbit with different orbital elements. This means that as the Earth moves around in the 3-body problem, its position and velocity at any moment define an elliptical orbit called an *osculating* ellipse (much like the osculating circle we saw in Chapter 3), which is the ellipse the Earth would follow if Jupiter were to vanish at that moment. The orbital elements of the osculating ellipses are called the *osculating elements*, so now we have 4 variables a, e, ϖ and v which all change with time; we need some differential equations to describe this change.

This program was carried out by (mostly French) mathematicians and astronomers, such as Clairaut, d'Alembert, Euler and Lagrange, culminating in Laplace's *Traité de Mécanique Céleste* - 5 volumes published between 1799 and 1825 ([73], see Figure 12.3). Crediting Lagrange, Laplace writes the differential equations for a planet in the form

$$\left(\begin{array}{c} \text{acceleration of} \\ \text{planet} \end{array} \right) = \left(\begin{array}{c} \text{force due to} \\ \text{the Sun} \end{array} \right) + \left(\begin{array}{c} \text{forces due to} \\ \text{the other planets} \end{array} \right)$$

where the forces due to the other planets are generated from a 'disturbing function', denoted R (see Figure 10.5). Now since the osculating elements of the planet can be written in term of its position and velocity, and vice versa, these equations can be recast in the form

$$\frac{da}{dt} = \dots \qquad \frac{de}{dt} = \dots \qquad \frac{d\varpi}{dt} = \dots \qquad \frac{dv}{dt} = \dots$$

(see Figure 10.5 for an example); these are known as *Lagrange's Planetary Equations*, and they are pretty nasty. To make progress, the disturbing function is expanded in a power series, much like the ones we have seen already, only now the terms in the series are sine's and cosine's of the various angles and linear combinations thereof. We then need to make a choice: the more terms we include in our power series the harder the equations will be to solve, but the more accurate our solutions will be. With this approach Laplace showed, in 1784 [91], that yes the orbit of Jupiter is getting smaller while that of Saturn is getting larger, but in time this will reverse: the variation in the semi-axes of these planets is *periodic*, just like our mass on the end of a spring bouncing backwards and forwards, only now the oscillation happens over a timescale of 900 years; pretty impressive! I would like to remind the reader that only 100 years had passed since Leibniz published the very first Calculus paper, in 1684, to the time Laplace showed, using Calculus, that subtle variations in astronomical observations were due to the complex gravitational interactions between the Sun and its planets that take place over centuries. But there was even better to come.

Mary Fairfax was born near Edinburgh, Scotland, in 1780 and married, becoming Mary Greig, but was widowed at 26. It was perhaps her widowed status that allowed Greig to move in scientific circles which ordinarily would not be considered appropriate for a woman at the time, and her mathematical ability and connections grew by submitting solutions to periodicals [108]. English mathematics was still stuck in a rut using the 'fluxional' calculus of Newton, and English scientists were looking with envy at the huge leaps forward on the continent. By the time Mary Greig married again, to become Mary Somerville, she was a master (mistress?) of the differential calculus of Leibniz. She met Laplace in 1817, and so impressed him the story went that Laplace claimed only three people understood his work: Ms Fairfax, Mrs Greig, and Mrs Somerville. She was commissioned to write an English translation of the *Mécanique Céleste* and in 1831 published *The Mechanism of the Heavens*, her clear writing style winning acclaim from experts as well as the broader scientific community (see Figure 10.5).

The accuracy of prediction brought by Laplace's *Mécanique Céleste* meant that even the slightest discrepancies between

Figure 10.4: Mary Somerville 1780-1872 [61]

observation and theory drew attention, in particular the orbit of Uranus did not quite match with the seven planet Mercury-to-Uranus solar system as it was known. In the 6th edition of her second book *On the connexion of the physcial sciences*, Somerville writes *If after the lapse of years the tables formed from a combination of numerous observations should be still inadequate to represent the motions of Uranus, the discrepancies may reveal the existence, nay, even the mass and orbit of a body placed for ever beyond the sphere of vision.* This sentence [106] inspired John Crouch Adams to search for an eighth planet using Lagrange's planetary equations; in 1845/6 both he and Urbain Le Verrier in France predicted the location of an eighth planet which was then observed by the Berlin Observatory. While the inevitable priority dispute followed as

Demonstration of La Grange's Theorem.

417. The equations which determine the real motion of m in its troubled orbit are, by article **347**,

$$\frac{d^2x}{dt^2} + \frac{\mu x}{r^3} = \left(\frac{dR}{dx}\right),$$

$$\frac{d^2y}{dt^2} + \frac{\mu y}{r^3} = \left(\frac{dR}{dy}\right), \qquad (87)$$

$$\frac{d^2z}{dt^2} + \frac{\mu z}{r^3} = \left(\frac{dR}{dz}\right).$$

If this value of de, and the preceding values of da, de, dp, dq, be substituted in equation (**113**), observing that $\dfrac{dR}{ndt}$ may be put for $\dfrac{dR}{de}$ and $\dfrac{dR}{d\varpi}$, it will be reduced to

$$d\varpi = \frac{andt \sqrt{1-e^2}}{e}\left(\frac{dR}{de}\right);$$

whence $de = \dfrac{andt \sqrt{1-e^2}}{e}\, (1 - \sqrt{1-e^2}) \cdot \left(\dfrac{dR}{de}\right) - 2a^2\left(\dfrac{dR}{da}\right)ndt.$

Figure 10.5: Excerpts from Mary Somerville's *Mechanism of the heavens*, 1831 [105].

to who should get the credit for the discovery, and who should name the planet, this is a good time to step back and take stock of what has been achieved: human beings, those hairless apes, developed in the blink of an evolutionary eye an elaborate scheme of scribbles, drawings and rules of manipulation, to such a degree that they were able to realize the existence of an entire planet that had been floating around the solar system for billions of years completely beyond direct observation. Mathematics was triumphant; this was the zenith of the Newtonian program and a brash confidence overtook the scientific community, it seemed like there was no problem that could not be solved by the new analysis. Laplace went so far as to imagine a vast intellect that knew at a certain moment all the positions and velocities of all the particles in the universe; if this intellect were vast enough to *submit these data to analysis, it would embrace in a single formula the movements of the greatest bodies of the universe and those of the tiniest atom; for such an intellect nothing would be uncertain and the future just like the past could be present before its eyes.* But as the 19th century passed to the 20th, this 'clockwork universe', predictable and Newtonian, was revealed to be as ephemeral as Ptolemy's epicycles, undone by two forces: chaos and curvature.

Henri Poincaré was one of the greats. Often described as 'the last universalist' because his research spanned all the mathematics of the time (an impossible ask nowadays), he made significant contributions to Geometry and Analysis, almost single-handedly established Topology and Dynamical Systems as branches of mathematics in their own right, and even wrote 'popular science' books for the general public. He is my own personal Mathematics hero, and perhaps the thing I admire most about his work was his creativity; he had the uncanny ability to approach a well established problem from a completely original direction, often expanding the topic into a larger abstract setting where wide-ranging powerful theorems and general principles revealed themselves, before moving on to the next subject. We could spend a whole chapter talking about the work of Henri Poincaré; oh well, Statement A.

And what better example of Poincaré's flair for the original than his work in celestial mechanics? In 1885, King Oscar II of Sweden and Norway established a prize (fame and acclaim, and 2,500 crowns) for the mathematician who could demonstrate the long-term stability of the solar system [32]. Though Laplace had shown the variations in the semi-axes of Jupiter and Saturn were periodic, he did so by ignoring 'small' terms in the power series as described previously; however, when the frequencies of the periodic motions are in ratio it is possible for these interactions to reinforce one another and grow without bound, a phenomenon known as 'resonance', and indeed there are many resonant arrangements in our solar system [85]. While the 19th century had been focused on Herculean calculations and extensive analysis, Poincaré favoured a more geometrical approach; he emphasized a *qualitative* rather than *quantitative* understanding of differential equations: rather than saying what the solutions *are*, we try and say what the solutions *do*. Poincaré's attack on the stability of the solar system was the best of mathematics: imagination paired with rigor, but even he stumbled, as we will see. The full treatment is quite technical, and we will present only a sketch of the big ideas here, which is nonetheless quite abstract; see what you think of it (I recommend [32] and [10] for the history, [67] and [66] for the technicalities).

Poincaré focused on the 3-body problem, and there are two jumping-off points: firstly, I remind you how in the last section we saw that the 'state' of the Earth (more generally we will just refer to a 'third body') was recorded by 4 numbers: either the position x, y and velocity $\dot{x}, \dot{y}$ or the osculating elements a, e, ϖ, v. These are not the only options, but whatever you choose you will always need to keep track of *four* variables; we call this 4-dimensional space 'phase space'. At any moment in time the state of the third body is given by values of these 4 numbers, which is a point in phase space; at a later time the state is 4 different numbers, which is another point in phase space. As we follow the state of the third body continuously through time it will sweep out a curve in 4-dimensional phase space (although for the pictures below we will draw it as 3-d). Since the state of the third body is dictated by some differential equations (such as those in Figure 10.5) we will refer to these curves as 'solution curves'. The qualitative viewpoint is to ask: what do these curves do? Do they, for example, go off to infinity? Or perhaps they get trapped in some region? Or tend to a single point?

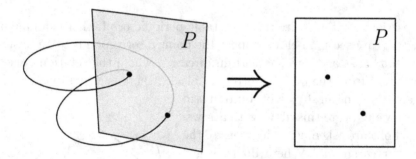

The second motif of his analysis is the 'periodic orbit': the path of the third body repeats itself, possibly after a long time or after doing some very complicated manoeuvres. In the phase space picture this means the solution curves are 'closed', as in the diagram on the left above. Because the dimension of phase space is high, Poincaré had the inspired idea to reduce the dimension by taking a plane in phase space and recording every time the solution curve intersects this plane (this is now called a 'Poincaré Section'). In the example above, the periodic orbit in phase space intersects the plane P twice and so the Poincaré section is just two dots (on the right).

Poincaré showed that periodic orbits come in two varieties: if we start at a point not exactly on the periodic orbit but just close to it, then we will either stay in the vicinity of the orbit (we call these 'stable') or drift away (we call these 'unstable'); in the image on the left below, the thick line is an unstable periodic orbit in phase space and the thin line is a neighboring solution curve that drifts away; in the Poincaré section on the right, the large dot is from the periodic orbit and the neighboring dots that drift away can be joined up to form an arc.

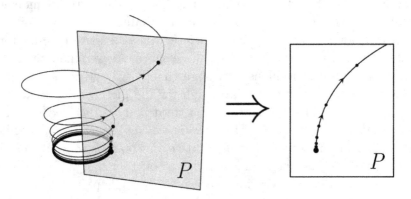

Of particular interest are these unstable periodic orbits. For example in the Poincaré section below suppose the point p corresponds to an unstable orbit. Just as there is a certain direction where the solution curves drift away from the periodic orbit (in black), there is also a direction where they naturally drift onto it (in gray); we indicate this 'flow' with arrows. Now Poincaré asked: what happens to the two curves shown? Where do they go? The natural option is to suppose they simply join up smoothly to form a loop, and that is exactly what Poincaré did; this closed curve separates the various regions in the plane of the Poincaré section,

and implies a form of long-term stability of the solar system (typical orbits are trapped in these loops and can't escape to infinity). Poincaré submitted his work to the competition, and he was duly awarded the prize in January 1889.

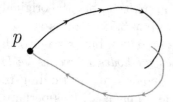

However, as his work was being pre-pared for publication, he noticed he had made an error, and quite a serious one: rather than the two arcs above sim-ply joining up to form a loop, it was in fact possible that they could *cross* one another. This completely changes everything. Remember the points on these arcs are where the so-lution curves intersect the plane; by the rules of the game, if these two curves cross once they must cross again and again, infinitely many times, and these crossing points get closer and closer as they ap-proach p. However, the area of each of the little regions between two successive crossings (light gray in the diagram on the right) must stay the same each time, which means the curves must stretch out and eventually start crossing one another yet again, in a terrifically complicated

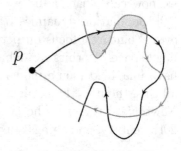

mesh-like structure; Poincaré said *one is struck by the complexity of this figure that I am not even attempting to draw* (but then he did score a 0 in the drawing element of his university entrance exams [32]).

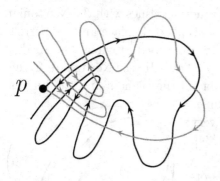

The implication of this struc-ture is dramatic: as the two curves mesh together closer and closer in a finer and finer web, points on one curve become infinitely close to points on the other curve. A random point chosen in this re-gion might lie on one curve and so the corresponding orbit would drift off in one direction, but an-other point infinitely close might lie on the other curve and so its orbit would drift off in a completely dif-ferent direction. This 'sensitive dependence' on the point chosen means the ultimate behaviour of the orbit of the third body is *unpredictable*, since you would need to know its position and velocity with infinite preci-sion. To be clear, there is no randomness here, no chance; the equations are completely deterministic, and yet the solutions are unpredictable. This was the death of the 'clockwork universe'.

Poincaré revised his submission to the competition, but the original version had already gone to print; he had to pay 3,585 crowns and 63 öre to have it destroyed (more than the prize money), but it was worth it: he ultimately published his results in *Les méthodes nouvelles de la mécanique céleste* in 1892, bringing a new depth of understanding to the complexity of the solar system never before imagined. It is perhaps likely that Poincaré and his contemporaries found this unpredictability distasteful, going against the zeitgeist as it did, and it wasn't until the arrival of the computer that mathematicians realized this complexity was hiding in plain sight: Poincaré's work was the first example of what we now call 'chaos', and we see chaotic structures in the solar system [12], population dynamics [75], geometry [124], the list goes on. Moreover, Poincaré's focus on periodic orbits gave us a way to organize the huge space of orbits available to spacecraft, something he could never have imagined (some modern applications are [58] and [121]). But for all this, the underlying physical framework of Laplace and Poincaré was Newtonian; this was the next sacred cow to be killed as we begin the 20th century, and things start to get weird.

What is a force anyway? How can the Earth feel a force from the Sun when they are not in contact? If Jupiter were to suddenly disappear, how would the Earth 'know'? And *when* would it know, instantaneously? There were several philosophical and practical issues with the Newtonian description of gravity, despite all the success of Laplace and Le Verrier. In particular, for Mercury, the angle ϖ described previously seemed to be growing at a rate larger than that predicted by the Newtonian framework, even when we take into account all the planets and the bulging of the Sun. Like a slowly rotating ellipse, this 'precession of perihelion' is very very small, 43 seconds of arc per century [34], but there nonetheless. Some proposed a new planet, Vulcan, inside Mercury's orbit, and some even said that rather than forces depending on the inverse square of distance, $\frac{1}{r^2}$ as Newton had proposed, they instead went like

$$\frac{1}{r^{2.0000001}}.$$

I hope the Pythagorean in you is disgusted. Poincaré said the solar system *may be subject to forces other than those of Newton* [32], and he was half-right: Einstein's bombshell was to say there is no such thing as gravitational force, inverse square or otherwise. Instead our universe is a four dimensional spacetime, and matter (like the Sun and the Earth, and you and me) *curves* spacetime around it, as we discussed in Chapter 3. Now when an object like Mercury tries to move in the curved spacetime surrounding the Sun its path must necessarily bend, and this is the effect we have for so long interpreted as 'force'. The intuitive way to picture this is as a rubber sheet with a bowling ball sitting in the middle as the Sun, and a tennis ball rolling across the sheet as Mercury. What path must it follow? For that we need the calculus of the next chapter.

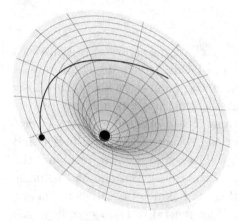

While Einstein introduced his Special Relativity (no gravity) in 1905, it took him 10 years to create a General Relativity (including gravity), and he was to some degree winging it: he tried one form of his 'field equations', saw did they work, scrapped them and tried something else, but all the while this precession of Mercury's perihelion was his guiding light. In his 1915 *Die Feldgleichungen der Gravitation*, Einstein writes [84] *the postulate of relativity in its most general formulation (which makes space-time coordinates into physically meaningless parameters) leads with compelling necessity to a very specific theory of gravitation that also explains the movement of the perihelion of Mercury.* It is this General Relativity which gave us black holes and big bangs, our current understanding of how the universe began, no less.

I gave a lecture once describing all of this, and a student asked me "what do *you* think is true? Is gravity really Newton's force or Einstein's curvature?" To which my slightly nihilistic response is: neither of them are true, any more than Eudoxus' spheres or Apollonius' epicycles; they are physical *models*, and all we can hope for is that some models are better than others. If you want truth, if you want certainty, then the only place you can get it is Mathematics. Of all the sciences, or even intellectual endeavours, Mathematics is special: it is the only realm in which you can prove something is absolutely, undoubtedly, irrefutably *true*, and always will be, even when the stars have stopped shining.

CHAPTER 11

Maxima and Minima

D IDO, THE STORY GOES, found herself on the shores of North Africa in the 9th century BCE, and the locals agreed she could buy as much land as she could contain with the hide of an ox. In a classic example of lateral thinking, she cut the hide into lots and lots of very thin strips, and formed from them a long thread of oxhide perhaps several miles long; she then used this to enclose a tract of land by the sea, and this is the founding story of Carthage (history does not record what the locals thought of her interpretation of the rules). But she had to make a decision: what shape should her thread of oxhide take so as to enclose the largest amount of land possible? More generally, 'Dido's problem' has come to mean: of all the closed curves in the plane of fixed length, which one contains the largest area? This is just one of many 'optimization' problems from ancient times, which you won't be surprised to hear were often phrased geometrically, for example: how to divide a line segment into two parts so the area of the rectangle formed by them is as large as possible? For example if you divide a segment so ⸻⊢⸻⸻ you make this rectangle ▮▮▮. The solution to this second problem was known to Euclid [86] and was a motivating example in the development of the derivative as we will see below, but Dido's problem was not formally solved until the 19th century and needed a whole new type of calculus called 'the Calculus of Variations' which then answered questions like: why are bubbles round? What is the shortest flight path from New York to Sydney? And how does the Earth orbit the Sun?

Perhaps the first person to find a maximum or minimum using what came to be known as the derivative was Pierre de Fermat, as early as 1629 (but not published until 1679 [109]). He begins with the classic problem:

    DOI: 10.1201/9781003455592-11

Sit recta A C, ita dividenda in E, ut rectang. A E C, fit maximum ; Recta A C, dicatur B.

A   E   C

ponatur par altera B, effe A, ergo reliqua erit B, — A, & rectang. fub fegmentis erit B, in A, — A² quod debet inueniri maximum. Ponatur rurfus pars altera ipfius B, effe A, + E, ergò reliqua erit B, —, A — E, & rectang. Sub. fegmentis erit B, in A, —, A² + B, in E, ²E in A, — E, quod debet adæquati fuperiori rectang. B, in A, — A², demptis communibus B, in E, adæquabitur A, in E⁴ + E⁵, & omnibus per E, divifis B, adæquabitur ²A + E, elidatur E, B, æquabitur ²A, igitur B, bifariam eft dividenda, ad folutionem propofiti, nec poteft generalior dari methodus.

Figure 11.1: A excerpt from Fermat's *Methodus ad disquirendam maximam et minimam* from 1679 [29]. Note his use of the word 'adequals'.

divide a line segment into two parts so the rectangle formed by them has maximum area (see Figure 11.1). If the total length is B, and one portion has length A and the other $B - A$, then the area will be $BA - A^2$. Fermat had read Kepler's musings on how the volume of a wine barrel changes with its proportions (wide versus tall). When tabulating the volume for incremental changes in proportion, Kepler observed *near the maximum, the decrements on both sides are initially imperceptible* [68], so Fermat said that the difference between the area at the maximum value of A and some nearby value, $A + E$, should be small when E is small, so he set them equal to one another, or rather he used the term 'adequal'. If we let $\approx$ denote 'adequals' then he is saying

$$BA - A^2 \approx B(A + E) - (A + E)^2.$$

Cancelling like terms he arrives at BE adequals $2AE + E^2$. Doing the old sleight of hand we saw in Chapter 9, he says: first E is not 0, so we can divide across by E to get $B \approx 2A + E$, then let E be equal to zero, so we get $B = 2A$. In other words, to make the area of the rectangle as large as possible you should divide the line segment at the midpoint, or we could say: if you have a fixed length of fence and you want to make a rectangle so it encloses the largest area, you should make it a square.

Fermat generalized: if you want the max/min of some expression depending on A, you adequal this to the same expression with A replaced by $A + E$ (or adequal their difference to zero), divide by E, and then let E be zero. Rephrasing in the notation of (9.3), we would write this as:

$$\text{set} \quad \frac{f(A + E) - f(A)}{E} \approx 0 \quad \text{then let} \quad E = 0,$$

which is pretty close to what we would say today: to find the max/min of a function, get the derivative and set it equal to zero. Fermat didn't seem to grasp that his method for finding maxima/minima was the same as for finding tangents, but Leibniz made the connection in his *Nova Methodus* (see the full title in Figure 9.2): when a function has a maximum or minimum then the tangent line to its graph is horizontal and therefore the slope of the tangent, which is the derivative of the function, is zero. Newton, inspired by Fermat's work, put it like this: *when a Quantity is the greatest or the least that it can be, at that moment it neither flows backward or forward... Wherefore find its Fluxion and suppose it to be nothing* [53]; we will therefore use the term 'stationary point' to refer to maxima or minima collectively. Let's look at an unusual example.

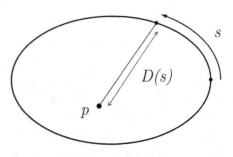

Suppose you have rowed out into a lake and you need to abandon ship and swim back to shore; which point on the shore is closest? Only a mathematician would say this, but let's suppose the lakeshore is an ellipse (on the right). If the point at which you find yourself in the water is p, and if we let s distinguish points on the ellipse (say $s = 0$ is the right-most point and as s increases we move anti-clockwise as the arrow suggests), then we can use Pythagoras theorem to write down a formula for the distance from p to any point on the ellipse and this will be a function of s; let's call it $D(s)$. When we draw the graph of $D(s)$ (Figure 11.2) we see there are four places where the tangent is horizontal and these four stationary points are therefore points on the shore which are nearer to or further from p than their neighbors.

We can say more, whatever the shape of the shore: if you follow the curve all the way round you come back to where you started (the curve is 'closed'), and if you were to keep going the function $D(s)$ would start repeating (it is 'periodic'). Since the portions where $D(s)$ is decreasing (dashed in Figure 11.2) and increasing (full) alternate, and there is a stationary point between any two such intervals, this means there must be an *even* number of stationary points, precisely as many minima as maxima. Notice also that if we were to draw circles centered on p of gradually larger radius, the circles would eventually touch the ellipse at the nearest point, and therefore the normal to the curve there (and at all the stationary points) would pass through p. It follows that there must

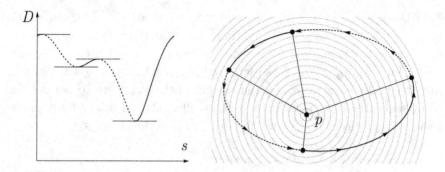

Figure 11.2: On the left is the graph of $D(s)$, with four stationary points. On the right we draw the line segments connecting p to those four stationary points, and those segments are then normal to the curve at those points.

be an even number of normals to the curve which pass through a typical point p!

Functions of a single variable like the previous examples are fine, but life is more complicated than that; what happens when we have a function that depends on several variables? This was again first considered through the framework of our old friend: curves in the plane. In the 1690s, Leibniz and Johann Bernoulli [53] studied families of curves, where one curve is distinguished from another by some parameter, let's say α. A motivating example is the family of lines formed by reflection off a circular mirror, like on the left below. Notice there seems to be a curve formed where these reflected lines focus; we will call it a 'caustic'. How can we capture this caustic?

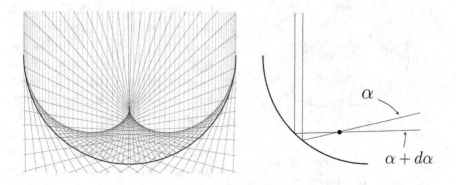

The equation of any one of those reflected lines would be something like $y = mx + c$, so y is a function of x, but now the coefficients m and c vary from one line to another and so they depend on α; as such y is a function of x *and* α, so let's write $y(x, \alpha)$. Suppose we take two infinitesimally close lines, say one has parameter α and the other $\alpha + d\alpha$. They will intersect when $y(x, \alpha + d\alpha) = y(x, \alpha)$, and in the limit as the two lines get closer and closer then the intersection is when

$$y(x, \alpha + d\alpha) - y(x, \alpha) = 0$$

as $d\alpha$ goes to zero. Leibniz [68] wrote this as $d_\alpha y = 0$ and called it 'differentiating from curve to curve', but now we would call $d_\alpha y$ the 'partial derivative' of y with respect to α. People initially just used the regular upright d for partial derivatives (like Somerville in Figure 10.5), but in the 1820s Jacobi introduced the ∂ symbol, so we would say the caustic of reflected lines is found from

$$\frac{\partial y}{\partial \alpha} = 0.$$

Next time you are having a cup of tea, look down at the surface and you will see a bright white line [93]: soon you will be seeing caustics everywhere.

Rather than reflected lines, another caustic of interest is that formed by normal lines, which is called an 'evolute', for example in Figure 11.3, we see an ellipse and several of its normal lines with the evolute becoming apparent; this is the same curve studied by Huygens that we mentioned in Chapter 3 in the context of the curvature of a curve. Notice the region inside the evolute is darker; this is because points inside this evolute have 4 normals passing through them, whereas points outside have only two (in fact we could *define* the evolute as the boundary between these regions [43]). This means the point p in Figure 11.2

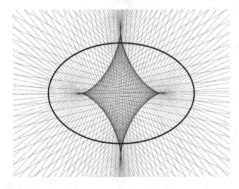

Figure 11.3: The evolute of an ellipse; note the spikes or 'cusps', a surprisingly ubiquitous feature [80].

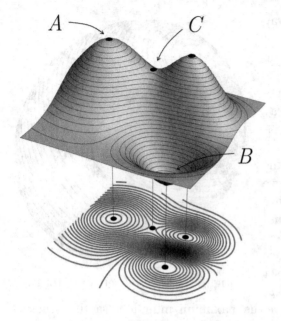

was inside the evolute of the ellipse.

As Calculus matured and moved beyond the geometrical to the analytical, mathematicians more generally thought of functions of several variables in the input/output sense: when we say $f(x, y)$ we mean you input two numbers, x and y, and the function outputs a third number, $f(x, y)$. This makes it natural to picture a function of two variables as a surface or sheet lying over the x-y plane, and its height above a certain point in that plane is $z = f(x, y)$. We said a function of a single variable was stationary when its derivative was zero, and so for a function of *two* variables to be stationary we need *both* its partial derivatives to be zero, i.e. both $\frac{\partial f}{\partial x} = 0$ and $\frac{\partial f}{\partial y} = 0$; this is good as it means we have two equations to solve for two unknowns, x and y.

As for the nature of stationary points, as before we have maxima (at A in the plot above) and minima (at B) but we also have a third possibility, the *saddle* (at C). Drawing pictures in 3D gets tricky, so instead we can take several horizontal planes and look at their intersection with the surface, and project those curves down into the x-y plane. These are called *level sets*, and another way to think of them is: all the points on a specific level set have x, y values that give the same output $f(x, y)$ (of course you have seen level sets before, as contour lines on physical maps or isobars on weather charts). Now look at the level sets below the stationary points: near the maxima and minima the level sets look like ellipses, whereas near the saddles they look like hyperbolae. In fact this is exactly how we test a stationary point to see what type it is, essentially using the classification Descartes and Fermat identified in the 1630s (see Figure 5.6).

This is something of a mathematical playground now - rather than a function defined on the plane like above, we can define a function on a surface; why not? The way to visualize this is with level sets, for example: extending the 'swim to the shore' idea from before (which was a function on a curve), let's say there is some point p in 3-dimensional space (the black dot in Figure 11.4) and some surface like a torus, and we define the function 'distance from p to a point on the torus'. Now we can draw the level sets of that function on the surface (all the points on a level set are the same distance from p)

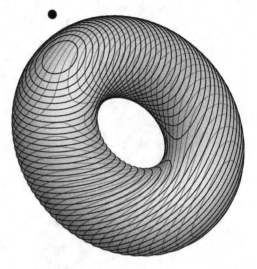

Figure 11.4: Level sets on a torus.

and we can see they still have the maximum/minimum/saddle structure near stationary points. In fact there is a restriction on the number of different types of stationary point we can have in this situation:

$$(\#\text{maxima}) + (\#\text{minima}) - (\#\text{saddles}) = 0.$$

For example in the Figure you can see a minimum and a saddle on this side of the torus, and a maximum and a saddle on the far side, and $1 + 1 - 2$ is indeed equal to 0 (to prove this we would need the ideas from Chapter 14). Everything extends naturally to functions of even more variables, but this is not the direction we want to go in; we want to extend beyond even functions themselves.

Nature is efficient. Nature is always trying to do the shortest, the quickest, the least, and this means we can formulate simple and universal principles that 'explain' natural phenomena. Two everyday examples from optics, Figure 11.5, are a good place to start: on the left, we see a ray of light reflecting off a surface. Euclid knew [86] that the two angles shown would be equal, and Heron (1st century CE) gave an explanation: off all the possible paths from A to B via the surface the one where the two angles are equal is the *shortest*. For proof, imagine the point B the same distance on the other side, at C; now the shortest path between A and C is a straight line (except of course for when it isn't, but we'll come back to that).

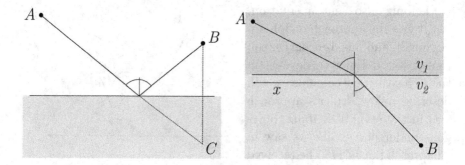

Figure 11.5: Reflection and refraction, nature being efficient.

More complicated is refraction, on the right in Figure 11.5. The speed of light in the upper region is different to the speed of light in the lower region, and Snel in the 1620s proposed a rule for determining the deflection of a ray of light as it passes from one region to the other, based on observations. Fermat gave the following explanation: of all the possible paths for the ray to get from A to B, the one that takes the *least time* is the one that satisfies Snell's law. Fermat's 'principle of least time', which also explained the reflection law, corrected a flawed attempt by Descartes, and this was the start of their bitter animosity [86]. For proof that Fermat's principle leads to Snell's law, we can write down an expression for the time taken for the ray of light to get from A to B as a function of x (see Figure 11.5), get the derivative and set it equal to zero; this is precisely what Leibniz did in 1682 (Fermat didn't quite yet have the calculus techniques needed). It seemed natural to look for other 'least time' type problems, and the next one lead to a whole new branch of Calculus.

In 1696, Johann Bernoulli posed a challenge to his fellow mathematicians: if a particle were to slide along a wire from A to B under its own gravity, what shape should the wire be so the time taken is the least possible? As a prize he offered *neither gold nor silver, [rather] we shall crown, honor and extol, publicly and privately, in letter and by word of mouth, the perspicacity of our great Apollo* [68]. He called this the 'brachistochrone' problem, and when Newton heard of the challenge he solved it in a single night; he sent his solution anonymously, but Bernoulli knew it was Newton straight away, saying *I recognize the lion by his paw* [86].

Leibniz also sent a solution, but it was the approach of Johann Bernoulli and his brother Jacob that inspired Euler to develop in the 1740s the new calculus we will describe below. Euler's approach was in the style of Leibniz, polygons under curves like we saw in Chapter 9, but in 1755 he received a letter from a then 19 year old Lagrange who proposed a whole new way of deriving the central

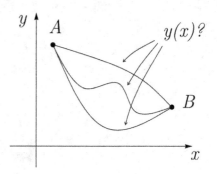

Figure 11.6: Set-up for the brachistochrone problem.

equation; Euler immediately recognized the superiority of Lagrange's formulation, even holding off publishing some of his own work so Lagrange could get full credit [16]. Every text, including this one, now uses Lagrange's approach, from which Euler coined the term 'Calculus of Variations', for reasons that will become clear.

If we look at the set-up for the brachistochrone problem in Figure 11.6, we see we are looking for some function, $y(x)$, describing the curve the particle slides down. If you made one choice for $y(x)$, giving a straight line say, then you can calculate a number for how long the particle takes to slide from A to B. If you make another choice for $y(x)$ (for example Galileo erroneously thought the curve should be an arc of a circle), then you get another number for the time taken. We could see it like this:

$$\left(\begin{array}{c} \text{choose a function } y(x) \\ \text{describing the curve} \end{array} \right) \rightarrow \left(\begin{array}{c} \text{find a number for how long it takes} \\ \text{the particle to slide along that curve} \end{array} \right).$$

This has the input/output form of function, only now the inputs are *themselves* functions; these 'functions of functions' are called *functionals*.

For some choices of $y(x)$ the time to slide from A to B will be large, and for some choices of $y(x)$ the time will be small; we are looking for the function $y(x)$ which makes the time *as small as possible*. Previously we saw that to find the max/min of a function you needed to get the derivative and set it equal to zero, so what is the analogous method for functionals?

For the brachistochrone problem, the formula for the time to slide from A to B is an integral, and the expression to be integrated depends on $y(x)$ and $y'(x)$, but it is the same for Dido's problem: the thing you are ultimately trying to minimize is an integral, so it makes sense to more generally look for a way of minimizing integrals like

$$J(y) = \int L(y(x), y'(x)) \; dx,$$

and we want to know how to choose y so $J(y)$ is a minimum. The term under the integral sign, denoted L, is called 'the Lagrangian', and its explicit form depends on the problem you are trying to solve: for the brachistochrone and Dido's problem it looks like [118]

$$L = \sqrt{\frac{1 + y'^2}{y}} \quad \text{and} \quad L = y\sqrt{1 - y'^2}$$

respectively. Lagrange said to suppose there was such a function y that minimized J, and to imagine small *variations* (hence the name of the technique) of the form $y + \epsilon\eta$. Just like Fermat, he said that near a minimum the difference between J evaluated for these two functions would be small when ϵ is small, i.e.

$$J(y + \epsilon\eta) - J(y) = 0.$$

Now using power series expansions Lagrange showed that if y is indeed a minimum of J then it must satisfy the equation

$$\frac{d}{dx}\left(\frac{\partial L}{\partial y'}\right) - \frac{\partial L}{\partial y} = 0. \tag{11.1}$$

This is the 'Euler-Lagrange equation', and it plays the role of $\frac{df}{dx} = 0$ for functions of a single variable. The algorithm is: find the L for your problem of interest, sub it into (11.1) which will give you a differential equation for y, and solve that equation for y; simples! The problem is the last step, as the equation you need to solve for y is often horrendous, but it's worth it: there are many problems of great importance that can be attacked successfully using this variational approach. We will look closely at two: mechanical systems, and geodesics.

For the record: the solution to Dido's problem is a circle, which you might have guessed (Weierstrass 1870s [118]), but the solution to the brachistochrone problem is the *cycloid*: the curve traced out by a point on a circle as it rolls along a line (next page). Intuitively this is because you want the particle to fall steeply at first to build up some momentum, before coasting to the end.

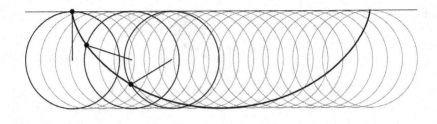

While Newton's second law, $F = ma$, was and continues to be one of the most important equations in Science, there were issues with the Newtonian approach both philosophical (as we saw in the last chapter) and practical: force and acceleration are vectors, and working with vectors is much more cumbersome than working with functions. For even simple mechanical problems we often need to introduce 'constraint' forces which complicates the picture, but also all Newton II does is give you the equations themselves; as for how to solve them you are on your own. For these reasons and others mechanics was completely reformulated by Lagrange in his *Mécanique Analytique*. Published in 1788, 101 years after Newton's *Principia*, Lagrange wanted to remove mechanics from the geometrical framework of Newton and rephrase it in the modern analytical language of calculus. To emphasize his break from Newton, Lagrange says *one will not find any figures in this work. The methods I present need neither constructions nor geometric. . . arguments but only algebraic operations* [98]. Lagrange showed that all the dynamics of a complex mechanical system, perhaps with many particles and constraints, could be contained in a single function derived from energies rather than forces. The approach was most succinctly expressed by Hamilton in his 1834 *On a general method in dynamics* [52]: if a particle has kinetic energy T (due to its motion) and potential energy V (due to the forces on it) then it will move from one point to another in such a way that the integral

$$\int (T - V)\, dt$$

is as small as possible (i.e. the Lagrangian function should be $L = T - V$). Integrals like this in the mechanical setting were called 'actions', and so this is known as Hamilton's *principle of least action*, and Calculus just swallowed Mechanics.

We will look at a complex example from the previous chapter where the advantages brought by Lagrange's reformulation of mechanics are crucial.

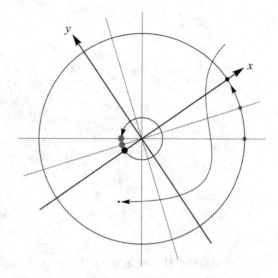

Poincaré in his *Les méthodes nouvelles* studied a particular type of 3-body problem: we suppose two of the bodies (called 'primaries') are large, like the Sun and Earth, whereas the third body is small enough (like a satellite) that it does not effect the orbits of the primaries, which are then supposed to be circular orbits about their center of mass. This is called the *circular restricted three body problem*, and the clever bit is the coordinate system used: we let the x-axis be the line joining and rotating with the two primaries, and the y-axis perpendicular to it, so in the x-y frame the two primaries are fixed (because the coordinate system is moving with them); all we need to worry about then is the movement of the third body.

We can derive some differential equations for the motion of the third body, using either Newton's approach or Lagrange's, you get the same equations either way, but you get something more with the latter: because the Lagrangian function does not depend explicitly on time, we can construct a *constant* which I will label C_J (the 'J' is for Jacobi), by which I mean a combination of the coordinates and velocities whose numerical value does not change as the third body moves around. You can measure the value of C_J at one moment, go away and come back in a million years, and when you measure C_J again it will have the same value. This does two things for us:

Firstly this constant C_J puts limits on where the third body can go. You can think of it like a 'total energy', and there are certain regions that the third body cannot reach because it doesn't have enough energy. In the picture below on the left we have a particular value for C_J and thus the shaded regions are inaccessible; this means the third body is either orbiting one primary or the other. On the other hand, the picture on the right is for another value for C_J and now the third body is able to pass from orbiting one primary to orbiting the other.

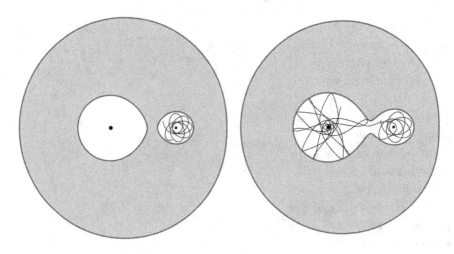

The second thing is deeper: these constants are called 'integrals', because they can be used to 'integrate' (i.e. solve) the equations, to the extent that Liouville showed in the 1850s [52] that if there are enough integrals then we can 'solve' the equations (apologies for all the inverted commas, it gets a bit meta when we dig into what it actually means to 'solve' an equation). For the 3-body problem Poincaré studied, you would need two integrals, but he showed in *Les méthodes nouvelles* that there is only one; as such we cannot solve the equations of the 3-body problem, which is ultimately why the behavior of the third body is unpredictable and indeed chaotic.

And why would there be integrals in the first place? Now we come to one of the deepest and most elegant theorems in all of Science: if there is a symmetry, there will be an integral, and vice versa. This is *Noether's theorem* which she published in 1918 in *Invariante Variationsprobleme* (see [52] or [120]). Broadly speaking, if there are enough symmetries of the mechanical system you are studying, and our use of the word 'symmetry' is slightly more abstract than the everyday, then there will be integrals and therefore you can solve the equations and the resulting dynamics will be regular and well behaved, like in the 2-body problem. On the other hand, if there is not enough symmetry and therefore not enough integrals, then we cannot solve the equations and the resulting dynamics will be irregular and chaotic, like in the 3-body problem.

Symmetry versus chaos, how very Greek.

To do geometry in the plane you need straight lines (how else can you define triangles, squares, distance?), and so if you want to do geometry on curved surfaces then you need something to play that role. *Geodesics* are the straight lines of curved spaces, and there are several ways of understanding geodesics because there are several properties of straight lines to generalize. The first would be that a straight line in the plane is the shortest path between two points, so we generalize this property as: of all the curves joining two nearby points p and q on a surface, the shortest one is a geodesic (see sketch on the right). We make this precise using the Calculus of Variations by writing

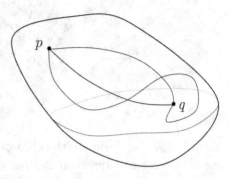

down a formula for the length of all the curves joining p and q, and then pick out the one that minimizes this length. If the length of a tiny element of a curve is ds, then we saw in Chapter 3 (equation (3.2)) how Gauss introduced the 'line element'

$$ds^2 = Edu^2 + 2Fdudv + Gdv^2$$

where u and v are a coordinate system on the surface. The total length of an arbitrary curve will then be

$$\left(\begin{array}{c} \text{length of} \\ \text{curve} \end{array} \right) = \left(\begin{array}{c} \text{sum of tiny} \\ \text{elements } ds \end{array} \right) = \int ds = \int \sqrt{Edu^2 + 2Fdudv + Gdv^2}$$

which we want to minimize; this tells us what our Lagrangian function should be and hence what differential equations we need to solve.

Some immediate examples: the geodesics of the plane are just straight lines, as you would expect of course, and the geodesics of a sphere are the 'great circles' as was implicitly understood by the Islamic scholars who developed spherical trigonometry all the way back in Chapter 1 (see Figure 1.6).

A more complicated example is the ellipsoid, which is like a sphere that has been stretched in three directions (see Figure 11.8). The equations for the geodesics are pretty nasty looking, but in

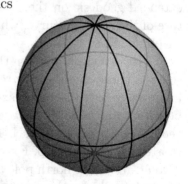

Figure 11.7: Some geodesics on the sphere.

Figure 11.8: Geodesics on the ellipsoid: on the left an arc of a geodesic as the shortest curve between two points, in the middle the same geodesic extended by moving 'straight ahead', and on the right a caustic of geodesics with four cusps clearly visible.

his *Vorlesungen über Dynamik* (based on lectures from 1842) Jacobi was able to 'integrate the ellipsoid', i.e. solve the geodesic equations [70], by finding the required integrals as described in the previous section.

A second property of straight lines is that they are, well, straight. If you were an ant living on the ellipsoid and I asked you to walk 'straight ahead' then you would be able to do so and your path would be a geodesic; we could say that the path you follow has no *intrinsic* curvature (as defined by Bonnet in 1848 [109]), even though to someone outside the surface it would look like your path is curved. This means we can follow geodesics as they wind their way around surfaces by just moving 'straight ahead'. For example in Figure 11.8 center, we see an extended geodesic on the ellipsoid; notice the regular well-behaved nature of the geodesic curve, which is due to the existence of the integrals Jacobi found.

There is a clash between these two properties: geodesics are only the shortest curves between *nearby* points. If you start at some point p and follow a geodesic to some nearby point q then it will be the shortest path between p and q, but if you were to keep moving straight ahead along the geodesic then you may reach a point where some other curve is now the shortest route back to p. For example look at the sphere in Figure 11.7: if you start at the north pole and start walking along a line of longitude it will initially be the shortest path back to the north pole, until you reach the south pole where there are now other shortest paths. In fact in

Figure 11.7 we can see several geodesics emanating from the north pole and meeting again, or 'focusing', at the south pole. In the same way, for surfaces in general we could take some point and imagine a whole spray of geodesics emanating from that point and observe where they focus, now not at another point but instead along a curve we will call a 'caustic' in analogy with our previous discussion. For example for the ellipsoid in Figure 11.8 on the right, the 'source' point is at the back and the caustic in front. In his *Vorlesungen* of 1842 Jacobi said the caustic on the ellipsoid has *die Gestalt der Evolute der Ellipse*, in other words it has four cusps and is 'circle-like' (as opposed to a figure-8 for example), see Figure 11.3. This became known as 'the last geometric statement of Jacobi' and it went unproved for 160 years. In 2004 Itoh and Kiyohara proved the caustic has four cusps [63]; that it is circle-like was proved in [126] (it gets really interesting in higher dimensions [128]).

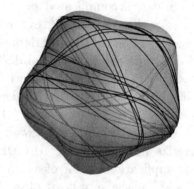

And don't be fooled by the pretty geodesics on the ellipsoid; in general, geodesics are just as chaotic and messy as gravitational problems. For example on the right we see a geodesic on a 'bumpy' sphere [124], and the scribbley erratic dynamic is clear to see.

But there is one last property of straight lines to mention: Newton's first law says that if a body is moving free from forces then its path is a straight line. This was just what Einstein needed to describe gravity without needing to invoke forces: if a body is moving in a curved spacetime, free from forces, its path is a geodesic. The classic phrase associated with Einstein's General Relativity is 'matter tells spacetime how to curve, and spacetime tells matter how to move', but now we know exactly how matter moves: on a geodesic. Right now you and I are moving along our own individual geodesics, falling forward in spacetime, along with everything else in the universe.

CHAPTER 12

PDEs

N ATURE IS CHANGE. You never step into the same river twice, and the way we capture that change is with the derivative. But the world is a complicated place, and we must therefore think about quantities that depend on several variables; how a quantity varies due to changes in one of those variables is captured by the partial derivative, and equations relating partial derivatives are called *partial differential equations* (PDEs). The differential equations we have seen so far have all involved quantities that depended only on a single variable, called *ordinary differential equations* (ODEs), and PDEs are more complex than ODEs in almost every way; in fact often our best strategy is to convert a PDE into (several) ODEs, as we will see. Due to their many many applications, some of which we will see in this and later chapters, PDEs are at the very heart of modern mathematics.

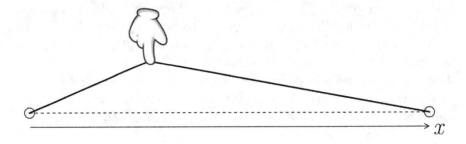

Figure 12.1: The plucked string.

Everything began with the vibrating string. Suppose a string under tension is pulled into the shape above and then released: what happens? The height of the string, h, above the dashed line varies as we

 DOI: 10.1201/9781003455592-12

move along the string, so h depends on x, but as time goes by the string is moving up and down so h also varies with time t. As such h depends on x and t, which we write as $h(x,t)$. In 1747 d'Alembert derived an equation for the height of the string which involved partial derivatives of h with respect to both x and t, the first *partial differential equation* (although elements of the derivation below were anticipated by Taylor in 1713 [109]). A short sketch would look like this: consider a small element of the string, between x and $x + dx$ [103].

Newton's second law relates the acceleration to the forces, so let's look at them one at a time: it seems reasonable to think that the higher I pull the string, the stronger will be the tension force; in other words, the tension depends on the slope $\partial h/\partial x$. For the element in the diagram, there are two tension forces: one due to the rest of the string to the left at x,

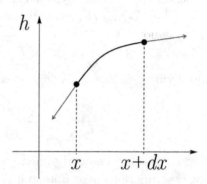

and one due to the rest of the string to the right at $x + dx$. They pull in opposite directions, so the total force is their difference:

$$\left(\frac{\partial h}{\partial x} \text{ at the right end } x + dx\right) - \left(\frac{\partial h}{\partial x} \text{ at the left end } x\right)$$

which is the *second* partial derivative with respect to x, written $\partial^2 h/\partial x^2$. The acceleration then goes like the second derivative of height with respect to time, written $\partial^2 h/\partial t^2$, so Newton's law says

$$\frac{\partial^2 h}{\partial x^2} = \frac{\partial^2 h}{\partial t^2}. \tag{12.1}$$

This is known as the *wave equation* or rather the 1-d wave equation (vibrations in 3-d are described by the 3-d wave equation in (12.6)). There are really some coefficients floating around but for the level of this text I am going to avoid them as much as possible by supposing they are all 1; indeed this is precisely what d'Alembert did when he solved the equation by an elaborate type of 'factorization' [109]. Two years later Euler published his own analysis along similar lines [68], but it is the work of Daniel Bernoulli (son of Johann) from 1753 that we want to dwell on; this next page is a little bit technical but stay with me, some big ideas follow.

We begin with what is known as an 'ansatz', which is a fancy way of saying you make a guess about the solution and see if it works. We suppose that the unknown function of x and t is really the product of two functions, one just of x and one just of t; let's write it as $h(x,t) = X(x)T(t)$ (we call this 'separation of variables'). Looking at (12.1) we need the partial derivative of h with respect to x, but all the x dependence has been put into $X(x)$ which is now just a function of a single variable, and so $\partial^2 h/\partial x^2 = X''(x)T(t)$ (the dashes here mean 'x derivative'). Similarly, all the t dependence is in $T(t)$ so we have $\partial^2 h/\partial t^2 = X(x)\ddot{T}(t)$ (the dots mean 't derivative'). Our wave equation (12.1) is now

$$X''(x)T(t) = X(x)\ddot{T}(t)$$

and dividing across by $X(x)T(t)$ we have

$$\frac{X''(x)}{X(x)} = \frac{\ddot{T}(t)}{T(t)}.$$

Now comes the classic argument: look at the fraction on the left. It only involves functions of x, and so it doesn't depend on t. But the left hand side is equal to the right hand side, so the right hand side also doesn't depend on t. Now look at the fraction on the right. It only involves functions of t, so it doesn't depend on x, so therefore neither does the fraction on the left. In other words, the left and right hand sides are both equal to the same constant. As any PDEs text will explain [56] we let this constant be $-k^2$ and we get two ordinary differential equations:

$$X''(x) = -k^2 X(x) \qquad \text{and} \qquad \ddot{T}(t) = -k^2 T(t).$$

We have seen this type of equation before, in (10.1): it is the equation of a spring, whose solutions are sine's and cosine's, but because the ends of the string are fixed and for other reasons we put the solutions of these two equations as $X = \sin(kx)$ and $T = \cos(kt)$, and therefore

$$h(x,t) = \sin(kx)\cos(kt).$$

As for the value of k, *any* positive integer will do, so we must include them all; we arrive at 'Bernoulli's solution' [103]:

$$h(x,t) = \alpha \sin(x)\cos(t) + \beta \sin(2x)\cos(2t) + \gamma \sin(3x)\cos(3t) + \dots \quad (12.2)$$

where the first term is when $k = 1$, the second is $k = 2$ and so on.

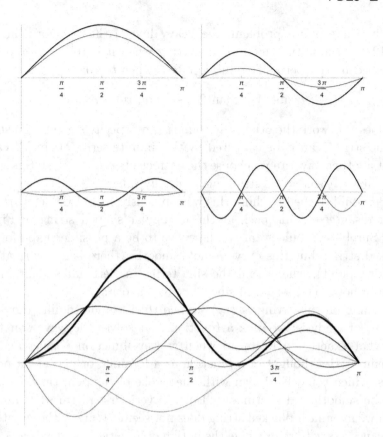

Figure 12.2: On top are the first four tones and below is the combination of those four tones; the heavy lines are the initial shapes.

Bernoulli called each term in this series a 'tone', where the $\sin(kx)$ part tells us the shape of the tone, and the $\cos(kt)$ part tells us how fast that tone is oscillating; the higher tones oscillate faster and therefore have a higher pitch, but the overall behavior of the string is a combination of these tones producing a rich and complex sound. In Figure 12.2, we see on the top the first four tones, and below a combination of them (we might call it a 'wave' or 'waveform'). In fainter lines we see the same tones after a fixed length of time; note how the higher tones have moved further than the lower tones, but also in each of the tones there are certain points which are fixed. These are called 'nodes' and are precisely as Pythagoras and his followers described: the length between one fixed end of the string and a node is a simple ratio of the whole length, and simple ratios produce harmonious sounds.

But there was a problem: the heavy line in the waveform in Figure 12.2 is the initial shape of the string when it is released, and it is described by the Bernoulli solution (12.2) when $t = 0$:

$$\alpha \sin(x) + \beta \sin(2x) + \gamma \sin(3x) + \ldots$$

but does this work the other way round? If we specify some initial shape for the string, can it be captured by this infinite series of sine's? Or to put it another way, can we choose the coefficients $\alpha, \beta, \gamma, \ldots$ so this series looks like *any* shape the string may initially take?

Bernoulli seemed to think this was obviously so, but d'Alembert and Euler disagreed. Look back at the triangular shaped string in Figure 12.1. Surely we would want to allow this to be a possible shape for the plucked string, but this curve is not 'smooth' (there is a corner at the plucking point), whereas all the sine terms in Bernoulli's solution *are* smooth; how could a sum of sine's not be smooth?

Pulling on this string (sorry) seemed to risk undermining the foundations of analysis: we seek a function that solves the wave equation, but what exactly do we mean when we say 'function'? Euler and his contemporaries thought of a function as a formula, or perhaps a power series, which could be drawn with one stroke of the pen; they assumed it to be smooth and continuous. But the very first picture we think of when we imagine a plucked string does not seem to satisfy these criteria. It was necessary to re-examine the notion of 'function', and this was part of the motivation behind the push for rigor in the 19th century.

How do you orbit a rubber duck? Or a giant space potato? Newton's *Principia* had treated gravitating bodies as perfect spheres, or (equivalently, as he showed) point particles, but nothing is a perfect sphere. On the right [26] is comet 67P/Churyumov–Gerasimenko which the Rosetta satellite got up close and personal to in 2014, and it is clearly not a sphere. Less dramatically, any round body that spins (like the Earth or the Sun) will bulge at its equator, and the challenge is to find the gravitational

forces acting on bodies orbiting such 'spheroids'. Laplace's approach in

PREMIÈRE PARTIE, LIVRE II. 137

mais on a

$$0 = \left(\frac{dd\,\delta}{d\,x^2}\right) + \left(\frac{dd\,\delta}{dy^2}\right) + \left(\frac{dd\,\delta}{dz^2}\right);$$

on aura donc pareillement

$$0 = \left(\frac{dd\,V}{d\,x^2}\right) + \left(\frac{d\,dV}{dy^2}\right) + \left(\frac{d\,dV}{dz^2}\right); \qquad (A)$$

cette équation remarquable nous sera de la plus grande utilité dans la théorie de la figure des corps célestes. On peut lui donner d'autres formes plus commodes dans diverses circonstances ; concevons, par exemple, que de l'origine des coordonnées, on mène au point attiré, un rayon que nous nommerons r; soit θ l'angle que ce

Figure 12.3: An excerpt from Laplace's *Traité de mécanique céleste* [73]; equation (A) is what has come to be known as 'Laplace's equation'.

his *Mécanique céleste* was to posit a function, called the 'potential function' which he denoted V, from which the gravitational force can be derived; it is easier to work with potentials rather than forces, as we saw in Chapter 11. Laplace then showed that the potential function must satisfy the following partial differential equation:

$$\frac{\partial^2 V}{\partial x^2} + \frac{\partial^2 V}{\partial y^2} + \frac{\partial^2 V}{\partial z^2} = 0, \qquad (12.3)$$

which is now called *Laplace's equation*, see Figure 12.3. This *équation remarquable* really has been terrifically useful, not just in gravitation but since Laplace's time it has been found to be of central importance in [46] electromagnetism, fluid dynamics, even in complex analysis, sadly none of which we will be able to discuss in detail, but I will use a feature of this equation to derive the last of the 'big 3 PDEs': the heat equation.

The sum of partial derivatives in Laplace's equation is called 'the Laplacian', so (12.3) could be read 'Laplacian of V equals zero'. The Laplacian has the following property first derived by Gauss [103]: it relates the value of a function at a point to the average value of that function at neighboring points. Suppose the function we are interested in is the temperature, v, which varies from point to point but also varies with time; as such v is a function of x, y, z and t. If the Laplacian of the temperature at a point p is positive, it means that neighboring points are *hotter* than at p, and so we would expect p to heat up: the temperature at p will *increase* over time. We could write it like this:

$$\left(\begin{array}{c} \text{Laplacian of } v \\ \text{at a point} \end{array} \right) \quad \text{is proportional to} \quad \left(\begin{array}{c} \text{rate of change of } v \\ \text{with respect to time} \end{array} \right)$$

or in symbols:

$$\frac{\partial^2 v}{\partial x^2} + \frac{\partial^2 v}{\partial y^2} + \frac{\partial^2 v}{\partial z^2} = \frac{\partial v}{\partial t} \tag{12.4}$$

(again I am suppressing constants of proportionality). This is the 'heat equation', first published in Fourier's *Théorie analytique de la chaleur* of 1822. In fact Fourier had submitted his analysis to the French Academy in 1807 [68] but it was rejected because of the same arguments that followed Bernoulli's solution to the wave equation: how to understand infinite trigonometric series. Fourier's approach to solving the heat equation was to suppose that a sufficient amount of time has passed by so the temperature has settled down to a 'steady state' which is not changing with time, in other words to set the time derivative on the right hand side of (12.4) to zero so we have again Laplace's equation (12.3), then do the same 'separation of variables' that Bernoulli had done to the wave equation. Faced again with the same series that Bernoulli encountered, Fourier made a detailed study of infinite trigonometric series which is why they are known today as 'Fourier series'; in particular when considering the steady state temperature in a semi-infinitely long rectangular sheet, he arrived at the following series:

$$\cos(x) - \frac{1}{3}\cos(3x) + \frac{1}{5}\cos(5x) - \frac{1}{7}\cos(7x) + \ldots \tag{12.5}$$

He noted that whenever he subbed in a value for x between $-\pi/2$ and $+\pi/2$, the series always summed to $\pi/4$, which is a bit odd: each individual term in the series is a function of x and thus varies as x varies; how could the sum then be a constant? And then whenever he subbed in a value for x between $\pi/2$ and $3\pi/2$ the series always summed to a *different* constant, this time $-\pi/4$, which seems odder still. As Fourier himself said: *these results seem to depart from the ordinary consequences of the calculus*. If we plot this series (see Figure 12.4), taking more and more terms, then we see the plot approaching the so-called 'square wave', which switches from positive to negative after fixed intervals. This seemed a million miles away from Euler's intuitive notion of 'function', and ultimately lead to a more abstract definition of function which we

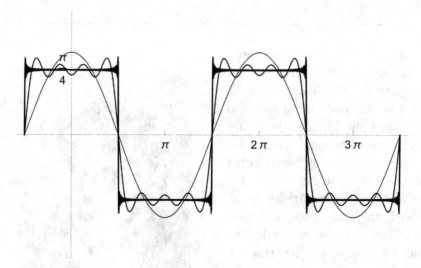

Figure 12.4: In progressively heavier lines, the series (12.5) with: the first term only, the first 4 terms, and the first 75 terms.

have used already in this text: simply a rule that relates inputs to outputs. In fact shortly after Fourier's work, Dirichlet [68] came up with the following monster: let $f(x)$ be 1 whenever x is rational, and -1 whenever x is irrational. This is perfectly fine as a function (for every input there is a well defined output), but it is *nowhere* continuous and *nowhere* smooth; I won't try and draw a picture.

The three PDEs we have seen so far are all called 'second order' because they involve second derivatives. The reason we include these three is that all second order PDEs fall into one of three classes, each with their own characteristics and quirks, and each of the 'big 3 PDEs' are representative of a particular class [64]; the naming convention harks back to Apollonius (Chapter 1):

Hyperbolic (Wave Equation) $\qquad \dfrac{\partial^2 h}{\partial x^2} + \dfrac{\partial^2 h}{\partial y^2} + \dfrac{\partial^2 h}{\partial z^2} = \dfrac{\partial^2 h}{\partial t^2} \qquad (12.6)$

Parabolic (Heat Equation) $\qquad \dfrac{\partial^2 v}{\partial x^2} + \dfrac{\partial^2 v}{\partial y^2} + \dfrac{\partial^2 v}{\partial z^2} = \dfrac{\partial v}{\partial t} \qquad (12.7)$

Elliptic (Laplace Equation) $\qquad \dfrac{\partial^2 V}{\partial x^2} + \dfrac{\partial^2 V}{\partial y^2} + \dfrac{\partial^2 V}{\partial z^2} = 0 \qquad (12.8)$

but already in Fourier's day there were some brave souls pushing forward into higher order PDEs.

Sophie Germain was born into a turbulent and revolutionary time; she was 13 when the Bastille fell and she and her family hid away in their Parisian home for safety, and she would while away the hours by reading books in her father's library. Euler caught her attention, but her parents disapproved; the story goes [113] that they kept her room dark and cold to discourage her reading, but she would smuggle in candles and blankets to study the master. When the École Polytechnic opened, Germain was not able to enrol due to her gender but she heard that one of Lagrange's students had dropped out of his course without telling anyone, so

Figure 12.5: Sophie Germain (1776-1831) [61]

she sent her own work to Lagrange under the borrowed name Antoine Le Blanc. Lagrange was impressed with Le Blanc's work and asked to see 'him'; Germain took a gamble and met with Lagrange, who was even more impressed and took Germain under his wing, introducing her to the faculty and encouraging her research. Her early work was in number theory, and she was responsible for renewing interest in, and making several contributions to, Fermat's last theorem (mentioned in Chapter 5). She wrote to Gauss, again using the *nom de plume* Monsieur Le Blanc, *fearing the ridicule attached to a female scientist* as she put it [41], but when her true identity was revealed Gauss wrote the following: *the taste for the ... mysteries of numbers is very rare ... but when a woman, because of her sex, our customs and prejudices, encounters infinitely more obstacles than men in familiarizing herself with their knotty problems, yet overcomes these fetters and penetrates that which is most hidden, she doubtless has the most noble courage, extraordinary talent, and superior genius.* High praise indeed (although he did go on to politely point out an error in one of her proofs). This was in 1807, and perhaps gave

Germain the confidence to tackle a mysterious physical phenomenon that was the talk of Paris.

In February 1809, Napoleon Bonaparte, Emperor of France, received Ernst Chladni at the Tuilerie palace. 20 years previously, Chladni had discovered that if he took a large metal plate, covered in fine sand, and dragged a violin bow across it to make it vibrate, the sand will immediately dance across the plate and settle down to make complex and beautiful patterns (see Figure 12.6 for Chladni's own drawings; below are my humble attempts with some electronic kit and table salt [93]).

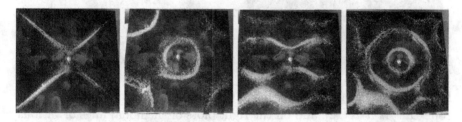

Napoleon was so intrigued he established a prize of 3000 francs, through the Paris Academy of Sciences, for the first to develop a mathematical theory explaining this phenomenon, and Germain was the only one who took up the challenge [113]. Her approach required her to consider the *curvature* of the plate as it vibrated; we have mentioned the principal curvatures previously in Chapter 3, and the Gauss curvature as their product, however Germain made particular use of the *sum* of the principal curvatures which is called the *mean curvature* although some authors refer to it as the 'Germain curvature'. Her work culminated in the following fourth order PDE:

$$\frac{\partial^4 z}{\partial x^4} + 2\frac{\partial^4 z}{\partial^2 x \partial^2 y} + \frac{\partial^4 z}{\partial y^4} + \frac{\partial^2 z}{\partial^2 t} = 0 \tag{12.9}$$

where z is the height of the plate over some point with coordinates x and y at time t. She was awarded the prize in 1816, the first woman to be so honoured; would you believe, she was not allowed to attend the award ceremony given that she was (gasp!) a woman. The Academy must have feared everyone would burst into flames if she walked into the room.

There are some who are less than complimentary of Germain's achievement: she was actually awarded the prize on her third attempt, with the first two being rejected due to several errors; the equation (12.9) was possibly first derived by Lagrange or Legendre; also her analysis left the question far from settled, and it was only in the following

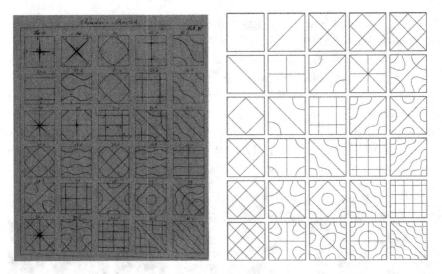

Figure 12.6: On the left are some of Chladni's original drawings from his 1802 *Die Akustik* [22], and on the right plots of the nodal lines of various tones due to analysis following from Germain's work [123].

decades that Kirchhoff and then Ritz [46] made further progress. But this view is quite uncharitable; Lagrange for example entered university as a teenager, and by 20 he was a professor of mathematics with the time and funds to do extensive research. Germain on the other hand was self-taught, having had a formal education not just denied her but actively discouraged; credit where it is due.

With the mathematical theory in place, it is now possible to make sense of Chladni's patterns. Look back to Figure 12.2; in general the vibration of a string is a complex shape but if we make it vibrate just right then all that we will see is one of the tones. Recall that each tone has certain points, called nodes or perhaps *nodal points*, where the string is not moving. The vibrating plate is the same: if we make if vibrate just right then we will see a 'tone', only now instead of nodal points there are *nodal lines* where the plate is not moving. The sand sprinkled on top of the plate will bounce around and naturally fall onto the nodal lines where there is no movement, and this is what we see in the Chladni patterns (see Figure 12.6).

Of the 'big 3 PDEs' you could argue the Laplace equation is the most fundamental, however it seems to me that the heat equation is the most ubiquitous. This is because heat spreads out: when you put a cold spoon in a hot cup of tea then the spoon gets hotter and the tea gets colder. This 'spreading out' is also called *diffusion*, and there are lots of things that are well modeled by diffusion. Just in my Maths department here in Portsmouth I have colleagues who are using the heat/diffusion equation to model electric car batteries [99], the evolution of dialects in languages [18], even the way coffee is brewed in an espresso machine [37]. We will see a final unexpected application of the heat equation in Chapter 16, but I want to finish this chapter with one last application of PDEs.

Imagine the Sun suddenly winked out of existence, how would we know? It would take 8 minutes for us to notice the absence of light, but what about gravitationally? Einstein's theory of General Relativity, which we have met several times now in Chapters 3, 10 and 11, says that matter curves spacetime, but how do changes in spacetime curvature propagate throughout the universe? Perhaps gravity is like heat, and spreads out? Or the universe vibrates like Chladni's plate?

The strategy is to think about small changes in spacetime curvature, derive an equation for how they evolve, and see if we recognize it; indeed this is precisely what Einstein did almost as soon as he arrived at his general theory, in 1916 [24]. We use the term 'metric', denoted g, to capture all the information about the shape of spacetime; we can think of the metric as a line element like in Chapter 3 or a matrix like in Chapter 8. Suppose g is initially something simple, and then give it a flick. We look at

$$g + \varepsilon h$$

where ε is very small; as such h contains the information about the small changes in spacetime due to some event. We take $g + \varepsilon h$ and push it through the field equations [34], remembering Fermat's trick that if ε is very small then ε^2 is very very small so we can ignore it. What pops out?

None other than the wave equation, equation (12.6). This means that gravity propagates like a wave, at the speed of light. Almost precisely 100 years after Einstein's prediction, gravitational waves were detected by the LIGO observatories in the US, in 2015 [24]. It is the stuff of science fiction: 1.3 billion years ago two black holes, each perhaps 30 times the mass of the Sun, orbiting one another 70 times per second, coalesced; the gravitational waves released by this event have been travelling across the universe ever since and were detected by measuring variations in length many thousands of times smaller than the nucleus of an atom.

The cartoon below gives an idea of the gravitational waves generated by two orbiting black holes.

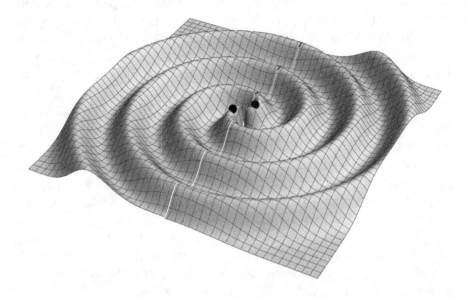

If you were to travel back in time to Archimedes or Euclid, and asked them what they hoped to achieve with the whole 'mathematics' thing, would they say "actually, I was hoping we could use maths to predict cosmic events of which we have no direct experience that happened billions of years ago on the other side of the universe"? How unreasonable! But this 'unreasonable effectiveness of mathematics' is behind the spectacular detection of gravitational waves, a miracle of engineering and physics but also a glorious triumph of mathematics. Because my friends: you can do Mathematics without Physics, but you cannot do Physics without Mathematics.

IV

Topology

CHAPTER 13

The Beginnings

W HEN ARE two things the same, and when are they different? Unfortunately humans are very good at detecting slight differences between one person or group of people and another, but we are also very good at overlooking superficial differences and seeing the essential characteristics of something: whether it is made of fallen logs or stepping stones, if it connects one river-bank to another we call it a 'bridge'. The image below [61] is (a 15th-century copy of) the first known map of Britain and Ireland, around 150 CE from the same Ptolemy we met in Chapter 10.

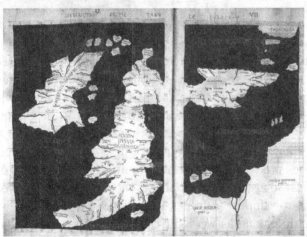

I doubt Ptolemy would have claimed this map was accurate from a 'metric' point of view, for example that the length of Ireland was truly in proportion to the length of Britain as the map suggests (plus I am not sure what Scotland is doing). Instead the map was valuable in a 'structural' sense: the land of the Eblanoi was connected to the land of the

DOI: 10.1201/9781003455592-13

Erdinoi but not to the land of the Trinobantes, and it was possible to sail up the west coast of Britain and come out in the North Sea. I am not saying Ptolemy invented Topology (or should it be Ptopology?), but Topology, like all Mathematics to some degree, takes these intuitive notions and weaponizes them, making them formal and precise and then pushing it as far as we can go. Arguably the first proper step in this direction was from Euler, in the 1730s (there are lots of books that introduce the ideas of this chapter, I highly recommend [96] which is itself a gem, also [42] and [21]). The city of Königsberg had seven bridges (on the left below), and the question was: is it possible to walk across all seven bridges without crossing the same bridge twice?

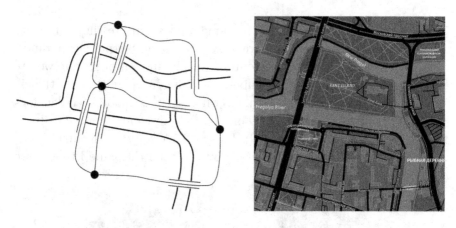

Euler described this as a problem of *geometriam situs*, the geometry of position, a term due to Leibniz although it is not clear what either of them meant. Euler's approach, completely novel, was to represent the problem as a 'graph' which we have met already in Chapter 8: each portion of land is a node, and each bridge is an edge connecting nodes, and Euler showed that the desired path is in fact not possible. The thing to note is that this has nothing to do with the lengths of the edges or the angles or areas between them; those geometrical properties are completely superfluous. Indeed we could stretch the edges or move the nodes around, as long as we don't change the connectivity of the graph the result holds, and this is one of the hallmarks of Topology as we will see. Funnily enough, since Euler's time bridges have been added and taken away (well it has been nearly 300 years) and now it *is* possible to walk across the five bridges of Kaliningrad, as Königsberg is now known, without crossing the same bridge twice (on the right above [20]).

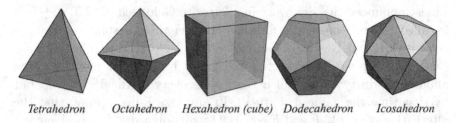

Tetrahedron   Octahedron   Hexahedron (cube)   Dodecahedron   Icosahedron

Figure 13.1: The five Platonic solids.

But it was another of Euler's contributions that was to be the archetype of a whole new branch of mathematics. We have seen examples of polygons before, such as triangles and quadrilaterals, and you will know that there are an infinite number of 'regular' polygons by which I mean one whose angles and sides are all the same, such as squares and hexagons. The 3D analogue of polygons are called 'polyhedra', but now there are only *five* regular polyhedra, shown in Figure 13.1. This result, originally due to Theaetetus around 400 BCE, took pride of place in Book XIII of Euclid's *Elements*. These five shapes occupied a central, almost mystical, place in the minds of mathematicians for centuries, including Kepler's as we saw in Chapter 10. Plato proposed that each of the four elements had as their atoms one of these regular polyhedra which therefore defined their nature; for example the atoms of earth are cubes which is why it is easy to build with earth, whereas the atoms of fire are spiky tetrahedra which is why it hurts when you put your hand in the fire. Plato saw these polyhedra as filled volumes, rather than hollow shells, which is why they are known as 'the Platonic solids'.

Well studied as they were over the ages, there is a very deep property of these polyhedra that no-one seemed to notice. Each polyhedron is made up of 2-dimensional polygonal areas like squares and triangles; these polygons meet one another along 1-dimensional line segments; and these line segments in turn meet at 0-dimensional points. We will call these 0-, 1- and 2-dimensional objects 'vertices', 'edges' and 'faces' respectively. For example look at the tetrahedron: it has 4 vertices, 6 edges and 4 faces, and we denote these numbers V, E and F. Now we calculate the alternating sum $V - E + F$ and we get 2. But whichever polyhedron you choose, whatever values they have for V, E and F, the alternating sum is always the same:

$$V - E + F = 2. \tag{13.1}$$

Euler announced this discovery in a letter to Goldbach in 1750 so (13.1) is called *Euler's polyhedron formula*, however there is some controversy since it turns out Descartes had made a similar observation in the 1630s (lost in a ship-wreck, recovered and noted by Leibniz, then forgotten about in a dusty corner till a chance discovery in the 1860s [96], but by that time the name had stuck). In any case Euler's proof, that the alternating sum of V, E and F equals 2 for all polyhedra, was problematic and several other proofs followed; the one we will give here is a variation on that of Thurston [116] which will be useful later.

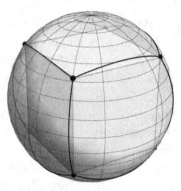

The first step is to project the polyhedron onto the sphere, as if a bright light is shining at the center of the polyhedron casting a shadow on the sphere; the vertices, edges and faces of the polyhedron then define vertices, edges and faces on the sphere, only now they are curved. (This was the first step of Legendre's proof, who then considered the areas of the faces on the sphere, using the methods we discussed in Chapter 3.)

This division of the surface of the sphere is called a 'triangulation', even though the faces need not be triangles. Having said that we can make them into triangles by simply adding edges joining vertices. An important observation is that if we do add an edge in this way then the sum $V - E + F$ stays the same: if you add an edge to divide a face in two, then the number of edges, E, goes up by one but the number of faces, F, also goes up by one and therefore $V - E + F$ simultaneously goes up by one and down by one, so stays the same. (This was a key step in Cauchy's proof, only he projected the polyhedron onto a plane rather than a sphere). The upshot is we can transform any one triangulation to another, they will all have the same value for $V - E + F$; we just need to show that value is 2.

So let's take some general triangulation of the sphere with some number of vertices V, edges E and faces F; a portion of such a triangulation is in Figure 13.2 on the left. Now Thurston says: suppose you have a bunch of particles, some with charge $+1$ and some with charge -1. We put a $+1$ charge at each vertex and in the middle of each face (black dots in Figure 13.2), and we put a -1 charge in the middle of each edge (white dots). The *total* charge will be

$$V \times (+1) + E \times (-1) + F \times (+1) = V - E + F. \qquad (13.2)$$

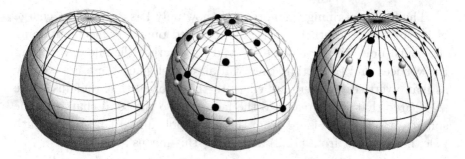

Figure 13.2: A portion of a triangulation of the sphere, with charges moving due to a 'flow' represented by the arrows on the sphere on the right.

Now we imagine a 'flow', like a current of a river, pushing the particles away from the north pole of the sphere and toward the south pole. As we 'flow' the particles we are not adding any or taking any away, and so

$$\left(\text{ total charge before flow } \right) = \left(\text{ total charge after flow } \right). \quad (13.3)$$

And what is the total charge after the flow?

If we have flowed the particles just a small amount, they will have moved off the edges and vertices and into faces. The full picture is a bit crowded, so in Figure 13.2 on the right we just show the central face: there are exactly as many $+1$ charges in it as there are -1 charges, and so the total charge in that face is zero. In fact this is the case for almost all of the faces: the total charge will be zero. There are only two exceptions: the face with the north pole, on the left below, and the face with the south pole, on the right (gray before the flow and black after).

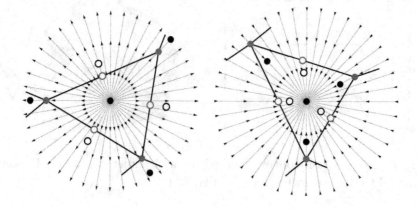

The face containing the north pole now only has a single +1 charge in it, since all the other charges have flowed out. The face containing the south pole will have some +1 and −1 charges that flowed in from the vertices and edges, but they will all cancel since there are exactly as many vertices as there are edges to the face; all that remains is the +1 charge that was already to the face, and so the total charge on the sphere after the flow is +1 from the north pole and +1 from the south pole, i.e. +2. But from (13.2) and (13.3) this means

$$V - E + F = 2.$$

So we have proved that for any polyhedron, the number of vertices minus the number of edges plus the number of faces is always 2. Except, of course, for when it's not.

Look at the object on the left of Figure 13.3, you probably have several in your house right now. It has vertices (16), edges (32) and faces (16), but now
$$V - E + F = 0,$$

not 2. The problem is Euler and many of those who followed him had things like the Platonic solids in mind when they thought of polyhedra, and those shapes are all 'convex', which means if we take any two points on the shape and join them with a line segment, then all the points on that line segment will be inside the shape. Euler's polyhedron formula, $V - E + F = 2$, does indeed hold for all convex polyhedra, and more besides, but not necessarily for *all* polyhedra.

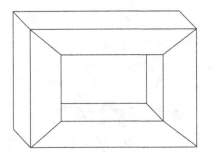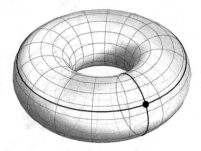

Figure 13.3: A non-convex polyhedron with Euler characteristic zero, inflated to make a torus (see also Figure 15.4).

Just like we could 'inflate' a cube to a sphere, we can inflate the picture frame to a torus (Figure 13.3 on the right), and the torus will then have vertices, edges and faces, only now $V - E + F$ will be equal to zero. This alternating sum of V, E and F for any surface is called the *Euler characteristic*, denoted χ, and now we can say things like

$$\chi(\text{sphere}) = 2 \quad \text{and} \quad \chi(\text{torus}) = 0.$$

As we argued before, the Euler characteristic does not depend on the choice of triangulation since we can go from one to another without changing χ; indeed in Figure 13.3 we see a triangulation of the torus which has 1 vertex, 2 edges and 1 face, and we have again $V - E + F = 0$ for the torus (we will contemplate this picture again in Figure 15.4).

You might think this is nothing more than a nice little combinatorial result, but it is quite liberating when you are only counting and not measuring. Suppose we take a sphere and put a triangulation on it, with a certain number of vertices, edges and faces. Now imagine the sphere is made of a very stretchy material, and we stretch and bend and twist the sphere into all sorts of odd shapes like in Figure 13.4. That stretching will carry with it the vertices, edges and faces, and their number will not change. This means $V - E + F$, the Euler characteristic, will not change; we say it is an 'invariant' (we will refine this below). However grossly you deform a sphere, you cannot change the Euler characteristic.

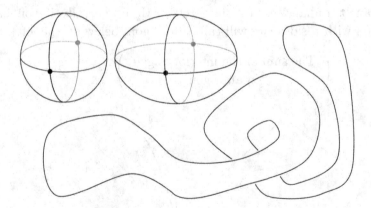

Figure 13.4: Some surfaces with Euler characteristic 2.

This means that you cannot stretch and deform a sphere to get a torus, because the torus has a *different* Euler characteristic; if you could deform one to another they would need to have the same χ. There is something fundamentally different about the sphere and the torus, deeper than their geometry, that the Euler characteristic is detecting, which then says something about the geometry. The finest expression of this interplay is, to my mind, the most beautiful equation in all of Mathematics: the global Gauss-Bonnet theorem. This theorem relates the Gauss curvature, K, of a surface S to its Euler characteristic, χ. As it is a special occasion, I will break with habit and give the equation in symbols before explaining it in words:

$$\iint_S K \, dA = 2\pi\chi(S). \qquad (13.4)$$

What makes an equation 'beautiful'? There is the aesthetic element: the elegant integral signs like the openings on a violin, the italic Roman mixed with the mysterious Greek; there is the conciseness: all the complex ideas and machinery developed over centuries, bound together in a short string of symbols; then there is the depth: the global Gauss-Bonnet theorem is a conduit, a wormhole, connecting two different branches of mathematics and making them one. That is the beauty; let us prove the truth.

We start with the theorem everyone learnt in school:

$$\left(\begin{array}{c} \text{The sum of the three interior} \\ \text{angles in a triangle} \end{array} \right) = \pi$$

(we say 'π radians' rather than 180°). More generally if you have a polygon with n sides (we will call it an n-gon, below on the left), then

$$\left(\begin{array}{c} \text{The sum of the interior angles} \\ \text{in an } n\text{-gon} \end{array} \right) = (n-2)\pi.$$

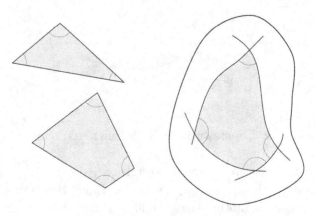

In Chapter 3 we met the Gauss curvature K, a way of capturing how curved a surface S is at any given point, and the *local* Gauss-Bonnet theorem (LGBT) for triangles drawn on the surface, which says: the sum of the angles in such a triangle is equal to π radians plus the Gauss curvature contained in that triangle (3.4). It generalizes easily to n-gons on S (see the figure above on the right):

$$\left(\begin{array}{c} \text{The sum of the interior angles} \\ \text{in an } n\text{-gon on a surface } S \end{array} \right) = (n-2)\pi + \left(\begin{array}{c} \text{The Gauss curvature} \\ \text{in the } n\text{-gon} \end{array} \right).$$

See how this equation contains the last two as special cases.

We now ask: what is the *total* Gauss curvature of the *whole* surface? By which I mean, we divide the surface up into some triangulation, use the LGBT to measure the Gauss curvature in each triangle, then add them all up. This is the approach described by Blaschke in 1921, but in fact the global theorem was first proved by von Dyck in 1888 [96]. If we use the Σ symbol for 'sum', we can write

$$\left(\begin{array}{c} \text{Total Gauss curvature} \\ \text{of the whole surface } S \end{array} \right) = \sum_{\text{faces}} \left(\begin{array}{c} \text{The Gauss curvature} \\ \text{in each } n\text{-gon face} \end{array} \right),$$

and then the LGBT above means we can write this as

$$\left(\begin{array}{c} \text{Total Gauss} \\ \text{curvature of } S \end{array} \right) = \sum_{\text{faces}} 2\pi - \sum_{\text{faces}} n\pi + \sum_{\text{faces}} \left(\begin{array}{c} \text{Sum of the interior} \\ \text{angles in each face} \end{array} \right).$$

That's the hard work done, let's look at each term on the right hand side one at a time.

The first term: the sum over all the faces of 2π just counts 2π for each face, of which there are F, so this adds up to $2\pi F$. The second term: n is the number of edges in each face, so the sum over all the faces of n just adds up all the edges in all the faces, but since every edge is in exactly two faces, one on each side, this counts each edge twice; so the second term adds up to $2\pi E$. The third term: adding all the angles in all the faces is the same as adding all the angles at all the vertices, but the sum of the angles at a vertex is 2π and there are V vertices, so the third term adds up to $2\pi V$. We have:

$$\left(\begin{array}{c} \text{Total Gauss} \\ \text{curvature of } S \end{array} \right) = 2\pi F - 2\pi E + 2\pi V = 2\pi \times \left(\begin{array}{c} \text{Euler characteristic} \\ \text{of } S \end{array} \right),$$

which is the global Gauss-Bonnet theorem.

The implications of this theorem are surprising. Suppose you were holding a balloon; you could draw some edges and vertices on it and calculate the Euler characteristic. The balloon is curved, perhaps in some places more curved than others. Now take your thumb and press it into the balloon at some point; this will grossly change the Gauss curvature at that point, however the *total* curvature of the balloon must stay the same, because just pressing your thumb into the balloon does not change the triangulation or the Euler characteristic, and from the global Gauss-Bonnet theorem we know this means the total Gauss curvature does not change. Squeeze and twist the balloon as much as you like, you cannot change the total Gauss curvature. Every time you think you have added some positive Gauss curvature somewhere, you must necessarily have added precisely the same amount of negative Gauss curvature somewhere else! All the surfaces in Figure 13.4 look different and have different Gauss curvatures in various places, but they all have precisely the same *total* Gauss curvature, namely, 4π. Moreover, not only does the torus have zero total Gauss curvature, because $\chi(\text{torus}) = 0$, but any shape you can make from a stretchy torus-shaped balloon will have zero total Gauss curvature, and any surface that has Euler characteristic zero, like those in Figure 13.5, will also have zero total Gauss curvature.

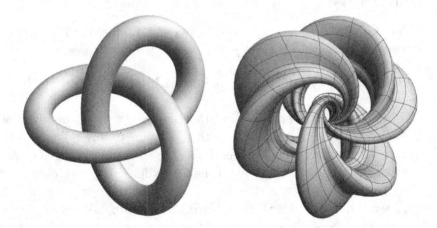

Figure 13.5: Some surfaces with Euler characteristic 0 and hence zero total Gauss curvature (the surface on the right is from [125]).

Perhaps you are getting the sense that there is something more, something deeper going on beneath this 'rubber sheet geometry', and to explore these depths we will need to get a bit abstract.

During the goldrush of the 19th century, this idea of deforming objects to discern structure beyond geometry was taken up by several prominent mathematicians [90, 110] and I will mention the main contributions in the developments of the next chapters, but it would be fair to say this was a disparate set of results without a coherent framework and terminology. That changed with the seminal work of Poincaré in 1895 in which he established Topology as a separate branch of mathematics in its own right. Ironically he didn't call it 'topology', he still used the term *analysis situs*, the analysis of position, a development of Leibniz's term *geometriam situs*. The problem with this term was it was very clumsy; what do you call someone who does analysis situs? An analysis situationist? In 1847 Listing had introduced the term 'topology' in his *Vorstudien zur Topologie* (since *topos* is Greek as *situs* is Latin), but it only really caught on in the early 20th century [96]. Based on our previous discussions, we could say Topology is concerned with the properties of objects which do not change as we deform them, and now we are able to be a little more precise about what that means.

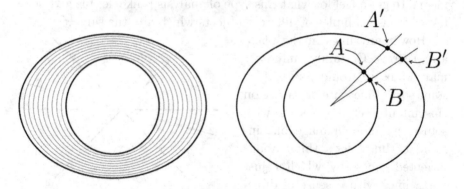

Figure 13.6: The circle and the ellipse are 'the same'.

Consider the circle and the ellipse in Figure 13.6. On the left we see how we can pass from the circle to the ellipse and vice versa via a family of intermediate curves; this chimes with the intuitive notion of 'continuous deformation' which we have already seen in Chapter 7. This is known as 'homotopy' (forgive me, I will avoid making technical definitions, but see [3] and [35] as excellent starters). While homotopy was already there in an intuitive sense in the work of Cauchy and Gauss as described in Chapter 6, it was really Poincaré who refined and developed the notion.

On the other hand, look at the circle and the ellipse on the right of Figure 13.6. By drawing a ray from the center outwards, we can identify the point on the circle that the ray intersects, A, with the corresponding intersection point on the ellipse, A', but also vice versa: each point on the ellipse A' has a corresponding point on the circle A. This correspondence is 'continuous', in the sense that if I take two 'nearby' points on the circle, A and B, their corresponding points on the ellipse A' and B' will also be 'nearby', and vice versa. This relation between the circle and the ellipse is known as a 'homeomorphism', a term coined by Poincaré, and is a little less intuitive in the sense that there need not be any intermediary curves between the circle and the ellipse so it is a little further from the 'rubber sheet geometry' picture. Nonetheless, homeomorphism is arguably the more fundamental notion in Topology, to the extent that if there is a homeomorphism between two objects we say they are 'the same' in a topological sense (for example the surfaces in Figure 13.5 are 'the same' as a torus). Depending on the setting we might be more interested in homotopy or homeomorphism, but in general I will try to be less formal and say things like two objects are 'the same' or 'topologically the same', or even they have 'the same topology' without clarifying which sense I mean. To get a feel for what this type of analysis is like, let me give some less obvious examples of different objects which are 'the same'.

How many lines in the plane are there? Certainly infinitely many, but we would like to give some sort of measure or handle on this infinite set. Let's start with something more manageable: instead of lines, let's think about 'oriented' lines, by which I just mean lines with a sense of direction on them denoted by an arrow.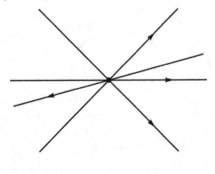
While two oriented lines may lie on top of one another, if they are oriented differently, one pointing one way and the other pointing the other way, then they are different oriented lines. Now let's ask: how many oriented lines in the plane are there that pass through the origin? We show some examples above.

We will call this set of oriented lines L_0, and the trick is to set up a correspondence between L_0 and some other set which we know already. To this end, let us consider a circle in the plane with center at the origin. Every oriented line through the origin will intersect the circle in exactly one place, if we only count the intersection *after* the line has passed through the origin (as we move along it according to its orientation).

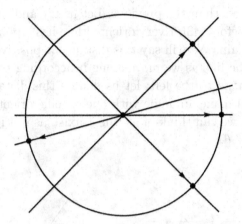

In the same way, if you choose any point on the circle then there is a unique oriented line that intersects the circle at that point, defining intersection as before. Therefore we have set up a correspondence between these two sets:

$$\left(\begin{array}{c} \text{Set of all oriented lines} \\ \text{through the origin} \end{array} \right) \quad \text{and} \quad \left(\begin{array}{c} \text{Set of all points} \\ \text{on a circle} \end{array} \right)$$

and this correspondence is continuous in the sense that if we take two 'nearby' oriented lines through the origin then they will correspond to two 'nearby' points on the circle, and vice versa. As such these two sets are 'the same', from a topological point of view.

This is some distance from the 'rubber sheet' picture we had before, so you might ask: what is the point? Well if these two sets are 'the same' then that means they share many properties; in 1911 Brouwer [110] showed that dimension is one of these shared properties. The circle is 1-dimensional, and therefore so is the set of all oriented lines through the origin of the plane. It is worth clarifying how we know the circle is 1-dimensional: we can use a parameter, or variable, to pick out a point on the circle, for example the angle, let's call it θ, that the radius from the center to that point makes with some fixed direction (although there are many other choices as we saw in Chapter 3). We say θ 'parameterizes' the circle, and since we only need one parameter then it is 1-dimensional.

Let's make this a little harder: how many oriented lines are there in the plane in general (not just those that pass through the origin)? Again there are of course infinitely many, but we expect this infinity to be somehow bigger than the previous. Let us try and parameterize this set, like we did before [43]. Every oriented line has a point on it which is closest to the origin. We will say this distance is positive if the origin is to the right of the line as we move along it according to its orientation, and negative if it is to the left; let us denote this distance d. Also the oriented line will make an angle with one specific oriented line through the origin which we can think of as 'the x-axis', as in Figure 13.7; let us denote this angle θ.

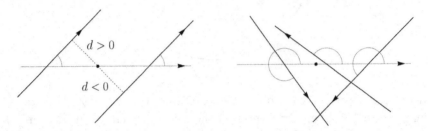

Figure 13.7: On the left we show two oriented lines that have the same θ but their d has different signs, and on the right we see several oriented lines with different θ (the horizontal line is our 'x-axis').

So we have two parameters capturing the set of oriented lines in the plane (let's call this set L). The parameter θ is an angle so it lives on the circle, whereas the parameter d is a distance, a real number, so it lives on the real line. Taken together, the two parameters describe combinations of circles and lines: for every point on the circle there is a line, and for every point on the line there is a circle. A circle of lines or a line of circles is a cylinder (see Figure 13.8).

Every oriented line in the plane defines a pair (θ, d) and hence a corresponding point on the cylinder. In the same way any point on the cylinder is identified by a pair (θ, d) which then corresponds to an oriented line in the plane. This correspondence is continuous: two nearby oriented lines correspond to two nearby points on the cylinder and vice versa; therefore these two sets

$$\left(\begin{array}{c} \text{Set of all oriented lines} \\ \text{in the plane} \end{array} \right) \quad \text{and} \quad \left(\begin{array}{c} \text{Set of points on} \\ \text{a cylinder} \end{array} \right)$$

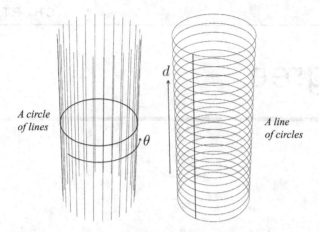

Figure 13.8: A circle of lines or a line of circles is a cylinder.

are 'the same'. Since we needed two parameters, the set of oriented lines in the plane L is 2-dimensional, unlike the set of oriented lines through the origin, L_0, which is 1-dimensional.

We now come to the central idea of Topology: we say X is an 'invariant' if any two objects which are topologically the same have the same X; alternatively we could say X is an invariant if it is 'conserved', or unchanged, by the continuous process that maps one object onto the other. An invariant might be a number, like dimension or the Euler characteristic; it might be a property, like whether the object in question is 'connected' (intuitively this means is it 'joined-up' rather than made of separate parts); or it might be something more sophisticated as we will see in later chapters. Topology is primarily the study of invariants, and we can use invariants to tell when two objects are topologically the same or not. For example we already saw that a sphere and a torus are not the same since they have different Euler characteristic, and neither of them are the same as the set L_0 since they have different dimension; might one of them be the same as L? And what about the set of all lines in the plane *without* orientation? Or the set of all planes in space?

Finally, if two objects that are topologically the same have the same invariant, is the reverse statement true: if two objects have the same invariant, must they be topologically the same? Sometimes the answer is 'yes', and a first example lies in the place where Topology began to crystallize in the mid-19th century: deforming curves in the plane.

CHAPTER 14

Degree

W HEN I WAS SMALL I would 'take a line for a walk': you draw a long loopy curve that crosses itself several times on a sheet of paper, making sure it ends at the same place it started, then color in all the little regions thus formed in different colors, and that is not a million miles away from some of the ideas and methods we will discuss in this chapter; if ever there were a case of Mathematics taking something intuitive and everyday, making it precise, and then blowing it up beyond all recognition, it would be this. The overarching concept is the 'degree of a map', and we will examine three manifestations which arise quite naturally in different contexts (although in [127] we managed to squeeze all three into one equation): first I will use the *winding number*, which we already saw in Chapter 6, to define the degree; then I will use the *rotation number* to classify plane curves via a 'complete' invariant; and finally I will use the *index* to study another of the great theorems of Topology, the Poincaré-Hopf index theorem, and make our first proper excursion into higher dimensional spaces. But we begin, as we so often do, with curves in the plane.

Let's take a closed plane curve, with an orientation (a sense of direction on the curve), and pick a point not on that curve; we will call it p, the 'base' point. We can draw a position vector from p to any point on the curve, and as we follow the curve around according to the orientation, the position vector will sweep around in the plane. Let's take a copy of the position vector and slide it so its tail is at the origin, and scale it so its length is 1; for example the points A, B and C on the curve on the left in Figure 14.1 correspond to the points A', B', C' on the unit circle on the right.

 DOI: 10.1201/9781003455592-14

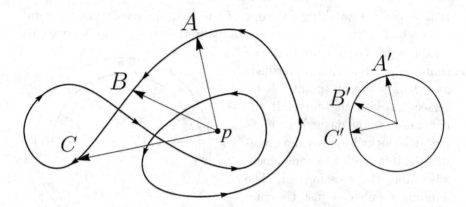

Figure 14.1: The winding number of the curve about the base point p.

Now as we move around the curve, this copy of the position vector will rotate around like the needle on a compass. When we have performed one complete circuit of the curve, the copy of the position vector will have completed a certain number of anti-clockwise rotations about the circle; we call this number the 'winding number' of the curve about the base point p. Intuitively it measures how many times the curve wraps around or encloses p, and the nice thing is that if we were to move the base point a little, or deform the curve itself a little, the winding number would not change. The fundamental reason for this is that the winding number is an integer, a whole number; we can make very small changes to p or to the curve, but the winding number cannot change by a small amount in response and so it does not change at all.

Let us rephrase this in a slightly more abstract way: for every point on the curve there is a corresponding point on the circle, and this is a *function* from the curve to the circle (the input is a point on the curve, like A, and the output is a point on the circle, A'). As we run through all the inputs of the function (i.e. we perform a circuit of the curve), we will cover the output of the function (i.e. rotate around the circle) a certain number of times, and this is called the 'degree' of the function (actually we tend to use the synonym 'map' in this setting, so we call it 'the degree of the map'). As such, the winding number is the degree of the 'unit position vector' map.

This is how most texts (for example [35] or [3]) introduce the degree, however the original definition of Cauchy in the 1830s [90] was quite different. A description closer to Cauchy's would be as follows [81]: suppose that as we are watching the copy of the position vector merrily rotate away as we allow its length to grow gradually thus tracing out a curve, like on the right. Draw a ray from the center outwards (shown dashed) and look at each point of intersection with the curve. If the curve crosses the dashed ray moving anti-clockwise label the crossing $+1$; if it crosses moving clockwise label the crossing -1. The winding number is just the sum of those $+1$'s and -1's. Notice if I chose another ray, say the undashed one on the right, then there might now be four points of intersection, but the extra two intersections would contribute a $+1$ and a -1 so the sum would be the same: 2.

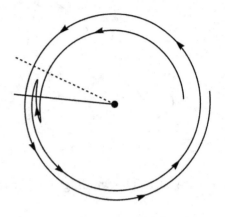

Cauchy introduced the winding number as a way of detecting roots of polynomials (he viewed the curve as lying in the complex plane, like we saw in Chapter 6), however nowadays it is usually mentioned in the same breath as a famous, or perhaps infamous, theorem from Topology: the Jordan curve theorem. This theorem states something that seems almost obvious, but turns out to be very hard to prove. Indeed while Jordan gave a proof in his 1887 *Cours d'Analyse*, most agree his proof was incomplete and that the first watertight proof was from Veblen in 1905; some authors now describe it as 'the most-proved theorem in topology' [3]. It says: if you have a simple closed curve in the plane (by 'simple' we mean the curve does not intersect itself), then this curve divides the plane into two regions; one region is bounded (we call it the 'inside') and one is unbounded (we call it the 'outside'). Seem true to you? Prove it!

The problem is that simple closed curves in the plane can be anything but; the obvious example is a circle, but a more elaborate example is given in Figure 14.2. The main argument of the standard proof is: since the curve is bounded, you can enclose it in a large enough circle; the winding number of the curve about some point on that circle is 0. Now as you gradually move that point toward the curve and at some stage cross it, the winding number will change and the thing to show is that as you cross the curve again and again the winding number goes from 0 to 1 and back again. If the winding number of the curve about some point is 0 we say that point is 'outside', and if it is 1 we say the point

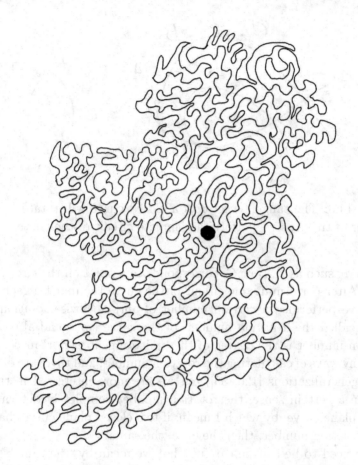

Figure 14.2: A simple closed curve in the plane. Is the black dot inside or outside?

is 'inside'. The black dot in Figure 14.2 is inside the curve, because in order to drag it to a place which we know for sure is outside we need to cross the curve an odd number of times.

The winding number depends not just on the curve but also on where we choose as base point; there is another 'turning' number which is just to do with the curve itself, called the *rotation number*. Beginning with an oriented closed curve in the plane, as we move around it then at any moment there is a sense of direction given by the tangent vector, see Figure 14.3. As before, we can drag each tangent vector so its tail is at the origin and scale so it has length one, so the tangent vector at any point on

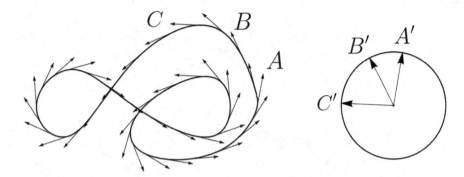

Figure 14.3: The tangent vector of a closed curve will rotate a certain number of times as we perform a complete circuit of the curve.

the curve, such as A, B or C, corresponds to a point on the circle, A', B' or C'. After performing one circuit of the curve the unit tangent vector will have performed a certain number of anti-clockwise rotations, and this is called the 'rotation number' of the curve (we could also say the rotation number is the degree of the 'unit tangent vector' map). There are many ways of calculating the rotation number, and there are of course many generalizations [126, 35], but we are more interested in the fact that, in a certain sense, the rotation number completely determines a closed plane curve, by which I mean: if two closed plane curves have the same rotation number, then they are 'the same'.

We need to be a little careful what we mean by 'the same': we are allowed to continuously deform and stretch the curve like an elastic band, but we are not allowed pull a loop 'tight', like in the diagram below.

Not allowed!

In 1937, Whitney [43] proved that if two closed plane curves have the same rotation number then you can deform one (bearing in mind what is allowed and what is not) until it sits on the other, with matching orientations. Thus not only is the rotation number an invariant (two curves which are the same have the same rotation number), but it is a 'complete' invariant (two curves which have the same rotation number are the same). This means we can use rotation number to classify curves;

for example all the curves in Figure 14.4 have rotation number 1, and so we can deform any one of these curves until it looks precisely like another.

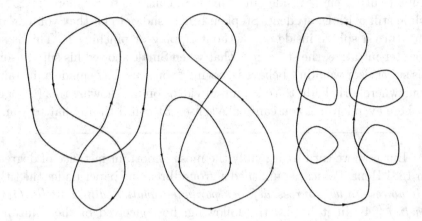

Figure 14.4: All of these curves are 'the same'; in fact they are all circles (with anti-clockwise orientation).

Think of it this way: we have a series of boxes. Into one box I put all the curves with rotation number 1, into another box all the curves with rotation number 2, and so on. All the curves in one box are the same, and if you give me a new curve all I need to do is calculate its rotation number and then I know which box to put it in. If you give me two curves, I can tell if they are the same or not by just checking do they go in the same box. This gives us a handle on *the whole set* of closed plane curves, all the ones that have been drawn before and all the ones that have not (see Figure 14.5).

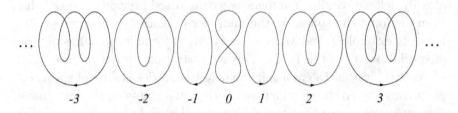

Figure 14.5: All the oriented closed plane curves there are, classified by rotation number.

Notice in Figure 14.5 there are two circles to either side of the central figure-8, one oriented clockwise and the other anti-clockwise. This means they have *different* rotation numbers, and so they are *not* the same; we cannot turn a circle inside out. This is in contrast to a sphere: in 1957, while still a Ph.D. student, Stephen Smale showed [43] that you can in fact turn a sphere inside out without ripping or pinching it. This is so counter-intuitive, the story goes that when Smale showed his supervisor, his supervisor wouldn't believe it, saying "you must have made a mistake somewhere!" He hadn't. You can find videos online showing this (Google 'sphere eversion'); if you can make sense of them, I tip my hat to you.

But now we turn to arguably the most important instance of degree. In 1881 Henri Poincaré published a groundbreaking paper under the title *Mémoire sur les courbes définies par une équation différentielle (1ére partie)* [94]. In it he did the following: he conceived of the solutions to differential equations as curves following a flow; he identified special points of the flow and invented a way to assign to them an integer called the 'index'; after projecting the flow to the sphere, he showed that the sum of the indices is 2.

Before we go any further, let's pause for a moment: surely any one of these original and far-reaching concepts is a 'big idea'; all three in one paper is taking the proverbial. How did he do it? Just as some people think the pyramids must have been built by aliens because humans aren't clever enough, in the same way I sometimes think Poincaré must have been visited by some time-travelling mathematician from the future who left him a copy of *Gray's Advanced Mathematics for the 21st century*. But the concept of flow was there in the fluid dynamics and electromagnetism of the mid 19th century; Cauchy and Kronecker had already defined winding numbers as we discussed previously; people had been mapping the sphere to the plane for centuries (just open an atlas for some examples); and Maxwell had derived formulae for sums of types of special points in the 1870s. This is not to take away from the vision of Poincaré, rather to emphasize that Mathematics is a team sport, one giant standing on the shoulders of another (it's giants all the way down!) Mathematicians *can* time-travel in a way, but only into the future; we can read today the words of Archimedes, Euler and Gauss, and it is like they walk among us still.

Let us unpack each of Poincaré's ideas one at a time.

*

We begin with something you have probably seen hundreds of times: a weather forecast.

The wind at any point has a direction and magnitude so we represent it with an arrow or vector like we saw previously, only now the wind varies from place to place; the way we capture that variation is with a vector whose elements depend on the x and y coordinates. We call this a 'vector field', a good example being

$$\begin{pmatrix} x^2 - y^2 \\ 2xy \end{pmatrix}. \tag{14.1}$$

At any point in the plane you feed in the x and y coordinates, then draw the resulting vector with its tail at that point; using the example above, at the point $(2, 1)$ (the gray dot in Figure 14.6), we would draw the vector $\begin{pmatrix} 3 \\ 4 \end{pmatrix}$. A better way to picture this is to imagine the path traced out by a particle as it follows these vectors around the plane, like a kite floating in the wind or a twig begin carried along in a stream; in Figure 14.6 the heavy line is the path that a particle starting at $(2, 1)$ would follow. Drawing several of these paths gives a sense of motion, and while there are various technical terms I will just refer to this as the 'flow'.

Poincaré's interest wasn't so much wind and water, but differential equations. While Lagrange

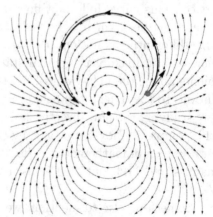

Figure 14.6: The flow due to the vector field given in (14.1).

had sought to excise Mechanics from the geometrical framework of Newton to make it a branch of Calculus (as we discussed in Chapter 10), Poincaré dragged Mechanics back into the world of Geometry, like putting on an old pair of well-broken-in shoes, only now with a more sophisticated abstract viewpoint. His justification was that too many simple mechanical problems lead to differential equations which cannot be 'solved' in a useful sense, so he instead proposed a qualitative approach: if we cannot say what the solutions *are*, we instead try to say what the solutions *do*. Let's take an example: the simple pendulum consists of a mass m at the end of a light rigid bar of length l which is fixed at one end but can otherwise move under gravity in the vertical plane (see Figure 14.7).

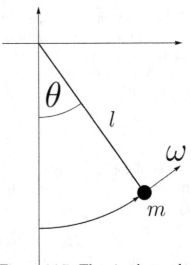

Figure 14.7: The simple pendulum.

Using the methods described in Chapter 10 or 11, we can derive a differential equation for the angle θ. Solve this equation and you know everything, but even in this simple instance that differential equation cannot be 'solved' in the usual sense. Instead, Poincaré said that at any moment in time the pendulum makes an angle θ and has velocity ω, but these variables change from one moment to the next so they trace out a curve in the θ, ω plane, called the 'phase space'. This curve follows a flow due to a vector field which is derived from the differential equation itself, and now we can make very general statements about the behaviour of the solutions to the differential equation, and hence of the pendulum, by simply studying the flow directly; no need to solve any equations! Arguably this completely novel approach by Poincaré was the beginning of a whole new branch of mathematics called Dynamical Systems or Nonlinear Dynamics [66, 112], but that, alas, is a story for another day (Statement A).

Something else to mention: we have met this phase space before. If θ is an angle it lives on the circle, and if ω is a velocity it lives on the real line. As such the phase space, being made up of pairs of values of θ and ω, circles and lines, is a cylinder (see Figure 13.8). The physical

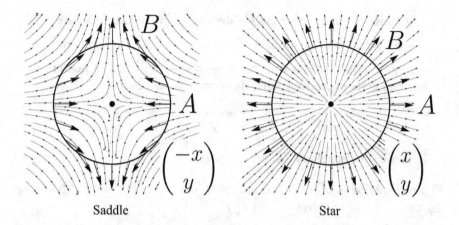

Saddle   Star

Figure 14.8: The zeroes of vector fields have special names, on the left is a 'saddle' and on the right a 'star', but there are also nodes, foci, centers... We also show (inset) the vector field that gives these flows.

make-up of the mechanical problem dictates the topology of the phase space which then in turn constrains the nature of the solutions, as we will see.

*

In the flows we have drawn so far, you will notice there are special points where the vector field vanishes (I will highlight them with black dots). I can think of at least four different names for these points from the literature, but in this text I will just call them 'zeroes'. They come in different varieties, and have special names (again these names vary from one place to another), and in Figure 14.8 we show two important examples. Poincaré defined an integer called the 'index' that we can associate to each zero in the following way: draw a circle around the zero, and at every point on the circle draw the vector from the flow that has its tail at that point. As we perform one complete anti-clockwise circuit of the circle, those vectors will have rotated anti-clockwise a certain number of times; this is the *index* (we could say the index is the degree of the 'vector field restricted to the circle' map). For the star (on the right of Figure 14.8), as we perform a single anti-clockwise circuit of the circle, the vectors will also have performed a single anti-clockwise rotation (for example in going from point A to point B the vectors in black are turning anti-clockwise); the index of a star is therefore $+1$.

Supposons qu'un point mobile décrive le cycle dans le sens positif et considérons les variations de l'expression $\frac{Y}{X}$. Soit h le nombre de fois que cette expression saute de $-\infty$ à $+\infty$; soit k le nombre de fois que cette expression saute de $+\infty$ à $-\infty$. Soit

$$i = \frac{h-k}{2},$$

le nombre i s'appellera l'*indice du cycle*.

Figure 14.9: Excerpt from Poincaré's *Mémoire sur les courbes* [94] defining the index of the vector field $\binom{X}{Y}$.

On the other hand, as we go around a saddle the vectors will complete a single clockwise rotation (Figure 14.8 on the left, where now as we go from point A to point B the vectors have rotated clockwise), and so the index of a saddle is -1. Actually Poincaré's original description (see Figure 14.9) was closer to Cauchy's definition of the winding number given previously (I paraphrase slightly): as we complete a circuit of the circle we observe each time the vector is pointing upwards, and assign a label $+1$ if it is moving anti-clockwise at that moment and -1 if it is moving clockwise; the index is the sum of the $+1$'s and -1's.

There are of course more precise ways of calculating the index (see [66] or [81]) using matrix determinants, but a special case is as follows: if the vector field is of the simple form

$$\begin{pmatrix} ax \\ by \end{pmatrix}, \tag{14.2}$$

then if $a \times b$ is positive the index is $+1$, whereas if $a \times b$ is negative it is -1 (for example, from Figure 14.8, the saddle has $a = -1$ and $b = +1$, so $ab = -1$ which is negative and therefore the index of a saddle is -1).

*

The problem with the plane is it is too big; it stretches out to infinity in all directions and so we can't see it all at once. One way to make the plane 'compact' is to project it onto a sphere, and here is the classic way of doing that (see Figure 14.10): you draw a line segment from the north pole of the sphere N to a point in the plane p; that line segment will intersect the sphere at a point q, and that point is then the projection onto the sphere of the point p. Since a curve is made up of a set of points, we can therefore project curves, and hence a flow, from the plane to the sphere and so we can consider flows on the sphere. A zero of the flow on the plane will be projected to a zero of the flow on the sphere, and they will have the same index.

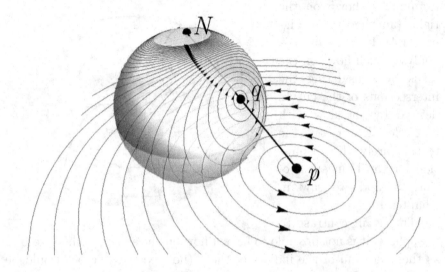

Figure 14.10: A flow in the plane projects to a flow on the sphere.

Poincaré said: *the sum of the indices of the flow on the sphere is 2.* This is known as 'Poincaré's theorem' to some, and it is a pretty strong statement: flows on the sphere can be very elaborate and complex,

an example from [127] is shown on the right; I also like to think of the swirling clouds of Jupiter with its vortexes and bands. Actually, we have already seen this theorem in action: in Figure 13.2 we described a flow that had a star at the north pole and a star at the south pole; both have index +1 as we know, and so the sum of the indices is 2. Why 2?

In a follow-up paper [90] a couple of years later Poincaré

extended this theorem: the sum of the indices of a flow on a surface is the Euler characteristic of the surface, and the Euler characteristic of a sphere is 2. But now we could imagine a flow on a torus for example, and since the Euler characteristic of a torus is 0 this means the sum of the indices must be 0. In particular, this means it is possible to have a flow on a torus with no zeroes at all, as Poincaré himself observed; an

example is shown on the right (another in is Figure 11.4). In fact Poincaré analysed such flows on the torus by considering the intersections of the curves defined by the flow with a circle drawn on the torus, surely the inspiration for the Poincaré Section method we saw in Chapter 10.

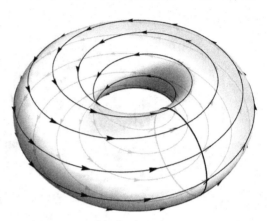

This is in contrast to the sphere: it is not possible to have a flow on the sphere with no zeroes, as then the sum of the indices of those (non-existent) zeroes would be zero, but we know the sum must be 2. This is famously known as the 'hairy ball theorem', because if you had a hairy ball (I am sure you are imagining a coconut or tennis ball) then each hair would be like a vector on the ball which would describe a flow, and that flow would have to have a zero (you couldn't comb the hair on the ball flat without a 'tuft' somewhere). Poincaré's proof was closer to Maxwell's 1870 *On hills and dales* [42], and nowadays the proof is often presented as a consequence of the Gauss-Bonnet theorem [35], but we are in the fortunate position that we have already prepared the ground to prove the theorem using Thurston's approach [116] from the last chapter:

Step 1 Suppose you have a flow on a surface; divide the surface up into faces, like we described in the last chapter, with the condition that each face either has one zero of the flow in it or none.

Step 2 Put a particle with charge $+1$ in the middle of each face and on each vertex, and a particle with charge -1 in the middle of each edge. The total charge therefore will be

$$V \times (+1) + E \times (-1) + F \times (+1) = V - E + F = \chi(\text{surface}).$$

Step 3 Allow the particles to move under the flow, just a little, so now all the charges are in faces. What is the total charge?

$$\left(\begin{array}{c}\text{Total charge after}\\\text{flowing particles}\end{array}\right)=\left(\begin{array}{c}\text{Charges in faces}\\\text{with no zeroes}\end{array}\right)+\left(\begin{array}{c}\text{Charges in faces}\\\text{with zeroes}\end{array}\right)$$

But the charges in faces with no zeroes all cancel; they make no contribution to the total charge (Figure 14.11 on the left). The charges that are in faces that do have a zero just add up to the index of that zero; an example is in Figure 14.11 on the right.

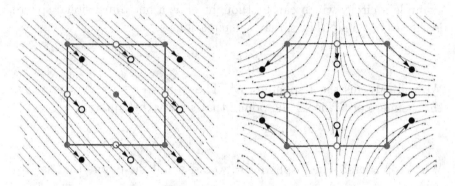

Figure 14.11: The positive charges are filled circles, negative charges empty circles; gray is before the flow, black is after. On the left is a face that does not contain a zero, and you can see there are two filled and two empty black circles inside it after flowing, so the sum of the charges is zero. On the right is a face containing a saddle, and after flowing there are two empty and one filled black circles, so the sum of the charges is -1, the index of a saddle.

Since the total charge is not changed when we allow the particles to move, the total charge before flowing equals the total charge after flowing and therefore

$$\left(\begin{array}{c}\text{Sum of the indices}\\\text{of all the zeroes}\end{array}\right)=\chi(\text{surface}).\qquad(14.3)$$

This theorem shows the importance of the topology of the space the flow lives on; it acts like a constraint, saying some things are allowed and some aren't. So on the sphere, you can't have a flow with no zeroes, try as you might. You could have a flow with one zero, but it would have to have index 2 (like the 'dipole' in Figure 14.6, but there are infinitely many ways to construct a zero so its index is 2 [42]); and if we did 'insert' a new zero into the flow somewhere, a saddle say, then we would necessarily be creating another zero somewhere else so the sum of the indices doesn't change - because it can't.

Spurred on by Riemann's manifolds (Chapter 3) and Hamilton's quaternions (Chapter 8), mathematicians of the late 19th century were emboldened to push geometry and algebra into higher dimensions, but it is hard to get a sense of what higher dimensional spaces are 'like'. For example a circle, which can be thought of as a one-dimensional sphere, is the set of points that satisfy the equation

$$x^2 + y^2 = 1.$$

We will call this the 1-sphere or $\mathbb{S}^1$. The 'sphere' we all know and love, the 2-sphere or $\mathbb{S}^2$, is all the points that satisfy the equation

$$x^2 + y^2 + z^2 = 1.$$

Why stop there? The 3-sphere, $\mathbb{S}^3$, would be all the points that satisfy

$$x^2 + y^2 + z^2 + w^2 = 1,$$

and so on to the n-sphere, $\mathbb{S}^n$. But these equations are a bit cold, and don't really tell us what the n-sphere is like. We can do the differential geometry of Chapters 3 and 11 and look at things like curvature and geodesics, giving us a sense of how round the n-sphere is, but these are to some extent 'local', they don't tell us about the *whole* sphere, and that is where topology comes in; indeed, in the early days topology was sometimes referred to as 'geometry in the large' since it captured this global structure. Let's focus on the 3-sphere, $\mathbb{S}^3$. In 1926 Hopf showed [90] that Poincaré's theorem (14.3) extended to *any* dimension, so we will use the index of Poincaré to calculate the Euler characteristic of the 3-sphere.

A nice way to get a handle on high dimensional spaces is to describe low dimensional spaces in a way that generalizes. For example, the 2-sphere on the left of Figure 14.12 has a flow going from the north pole to the south pole. Imagine we cut the 2-sphere along the equator to create two hemispheres; we are allowed to cut things as long as we remember to glue them together again, and the way we do this is to 'identify' points on both sides of the cut. So if a particle was moving away from the north pole due to the flow on the northern hemisphere and reached the equator, it would simply reappear on the southern hemisphere and carry on to the south pole, as if there were no cut at all. We can flatten the two hemispheres to get two 'discs', which is the region inside a circle, and these discs have a flow on them and they each have one zero: on one disc there is a star with the flow pointing outwards, and on the other a star with the flow pointing inwards.

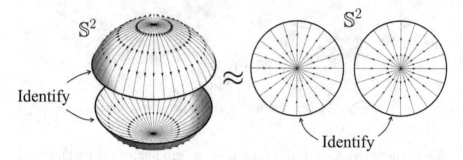

Figure 14.12: One way to visualize the 2-sphere.

The outwards flow is due to the vector field $\left(\begin{smallmatrix} x \\ y \end{smallmatrix}\right)$ (see Figure 14.8), and looking at (14.2) the product of the coefficients of x and y is positive so the index is $+1$. The inwards flow is due to the vector field $\left(\begin{smallmatrix} -x \\ -y \end{smallmatrix}\right)$, and the product of the coefficients is again positive so the index is again $+1$. By Poincaré's theorem the sum of the indicies is the Euler characteristic, so we have

$$\chi(\mathbb{S}^2) = +1 + 1 = 2,$$

which we already knew. Here's the payoff: we take the statement 'the 2-sphere is what you get when you take the interior of two 1-spheres and identify their boundaries', and just add 1 to each of the dimensions: *the 3-sphere is what you get when you take the interior of two 2-spheres and identify their boundaries*, see Figure 14.13.

Now there is again a flow in each region, and if a particle were following the flow on the left then when it reached the boundary it would simply reappear in the region on the right and carry on following the flow. The flow on the left has a zero which is a star with all the arrows pointing outwards, so it is due to the vector field $\left(\begin{smallmatrix} x \\ y \\ z \end{smallmatrix}\right)$. The coefficients of x, y and z are all $+1$ so a natural extension of the rule from (14.2) is that since the product of the coefficients is positive then the index is $+1$. Now we get the change: the flow on the right has an inward pointing star,

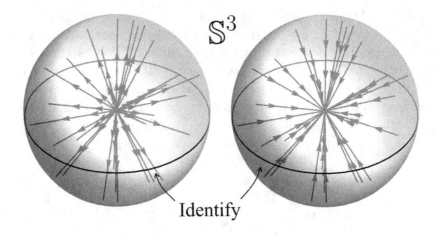

Figure 14.13: One way to visualize the 3-sphere.

due to the vector field $\begin{pmatrix} -x \\ -y \\ -z \end{pmatrix}$, and now the product of the coefficients is *negative*, so the index is -1. As such the Poincaré-Hopf theorem says

$$\chi(\mathbb{S}^3) = (\text{ sum of the indicies }) = +1 - 1 = 0.$$

This means that even though we drew a vector field in the 3-sphere which had two zeroes, it is in fact possible to draw one that has *no* zeroes; you can't comb the hair on a 2-sphere flat, but you can comb flat the hair on a hairy 3-sphere.

Now we have done the hard work, we can keep going with very little effort: a 4-sphere is harder to visualize, but it can again be thought of as the interior of two 3-spheres with their boundaries identified. We could give it a flow which has two zeroes, again one outward pointing star and one inward pointing star, with vector fields

$$\begin{pmatrix} x \\ y \\ z \\ w \end{pmatrix} \quad \text{and} \quad \begin{pmatrix} -x \\ -y \\ -z \\ -w \end{pmatrix},$$

whose indices are now both $+1$ again, so we can see

$$\chi(\mathbb{S}^4) = +1 + 1 = 2.$$

In fact it goes on alternating like this: the n-sphere has Euler characteristic 2 if n is even, and 0 if n is odd; a hairy 18-sphere would have to have a 'tuft' somewhere, but we could comb flat a hairy 19-sphere!

CHAPTER 15

Homology

N OW THAT WE HAVE gotten to know each other, it is time to open the fine wine. Topology came of age in 1895 with Henri Poincaré's *Analysis Situs* [95] where he introduced the notion of *homology*, which can be loosely described as using Algebra to count the holes in an object. The problem with this description, and the reason people can't agree on how many holes a straw has, is that the word 'hole' is very ambiguous; so before we describe homology, we will do a little hole-ology.

When we introduced the Euler characteristic, we pointed out that since the sphere and the torus have different characteristic this means we cannot deform one until it looks like the other; they are fundamentally different. What is this fundamental difference? Almost everyone will immediately say: it is that the torus has a hole, whereas a sphere does not. But a sphere *does* have a hole: have you ever inflated a football? Where is the air going? The hole in the middle. If spheres didn't have holes there would be no balloons, or jammy doughnuts, or profiteroles.

Now you might say: well a sphere doesn't have a hole in its *surface*, which is why we can have balloons that don't leak. But let's take a sphere with a hole in that sense, like you have pricked a hole in a balloon to deflate it (Figure 15.1 on the left). If we allow continuous deformations of a surface and consider anything we arrive at to be 'the same' as our original, then the sphere-with-a-hole is the same as a plate or a bowl (more properly a 'disc', Figure 15.1 on the right). Does a plate have a hole in it? Or a bowl? No-one would be able to eat soup. But if the sphere-with-a-hole is the same as a plate, how can one have a hole and the other one not? Does your sock have a hole? I hope so, otherwise how will you put your foot in it? Every bucket needs holes, so you can carry it. And if you add a hole to a sphere, so you get something which has no holes, does that mean [132] that the sphere originally had −1 holes? We seem to be unable to decide if a sphere has 0 holes, 1 hole or −1 holes, and this is largely because there are different *types* of holes, of different dimension, but also the whole point of a hole is that it is the *absence* of something. This leads to two questions: how do you count things that aren't there? And why would you want to?

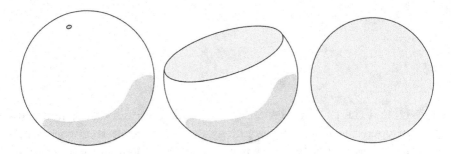

Figure 15.1: Is a sphere with a hole the same as a disc?

The second question was motivated primarily by Riemann, who in the 1850s [109] devised a completely novel method for understanding functions of a complex variable (in Chapter 6 we thought of a curve in the complex plane as an input, being mapped to another curve as the output; Riemann thought of the whole complex plane as the input leading to a bizarre object, with holes, as the output). As for the first question, a hint is in the Figure above: to give you a sense of the pierced hole in the sphere on the left, I drew the circular *boundary* of the hole. It was primarily Betti, building on Riemann's ideas, in the 1870s [68] who realized the significance of boundaries in detecting holes, and when Poincaré described a formal method for detecting holes in his *Analysis situs* via a series of integers he called them *Betti numbers*.

While homology and Betti numbers are typically described using group theory in modern texts [7], this was not the approach Poincaré took in *Analysis situs*, even though he referred to the notion of group often and indeed introduced there the *fundamental group* (which we explored in Chapter 7). Instead, he said that if ν_1 and ν_2 are the boundaries of a region then this defines a 'homology relation', written

$$\nu_1 + \nu_2 \sim 0$$

(where we read ~ 0 as 'is homologous to zero') and these relations can be combined like ordinary linear equations, much like we saw in Chapter 8.

We will therefore follow Poincaré's approach and use elementary Linear Algebra methods to compute homology directly as a way of introducing the main ideas, terminology and notation, and as such you will perhaps find this chapter different to the others in this book. I will try as usual to keep the jargon to a minimum, but if we want to see some non-trivial computations of homology then we need a little formality; I encourage you to stick with it though, this is the good stuff.

It is my firm, and tested, belief that if you want to understand something complicated then it is a good idea to start with a simple example and gradually make it more complex. To that end, consider the object below, with the vertices (x and y), edges (a, b and c) and faces (A) labeled; note also we use arrows to put an (arbitrary) orientation on the edges and the face. Words like 'curve' and 'surface' don't seem general enough any more so I will use the word 'manifold', and label it $\mathcal{M}$.

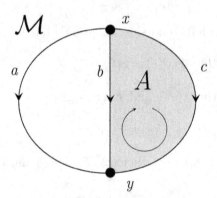

We can see there are many loops in $\mathcal{M}$, for example

$$a - b + c - a$$

is a loop, since it starts and ends in the same place, namely, x. Other loops are

$$b - c, \qquad -a + b, \qquad c - a + c - a$$

and so on. In fact there are infinitely many loops, but they are all combinations of a small set of loops, called a *basis*. The basis will have a certain size, and this is the *dimension* of the set of loops. Let's calculate this first, refining slightly the methods of Chapter 8.

The trick is to think of each edge as having a *boundary*, which are just vertices. We use the symbol ∂ for 'boundary', and we write the boundary of an edge as 'terminal vertex minus initial vertex'. For example since the edge a comes out of x and goes into y we would write

$$\partial(a) = y - x, \qquad \text{or} \qquad \text{'the boundary of } a \text{ is } y - x\text{'.}$$

Similarly, $\partial(b) = y - x$ and $\partial(c) = y - x$. A loop will just be some combination of edges which starts and ends at the *same* vertex, or in other words, a combination of edges with zero boundary; I know $b - c$ is a loop since its boundary is

$$\partial(b - c) = \partial(b) - \partial(c) = (y - x) - (y - x) = 0.$$

To find the set of *all* loops, we get the most general combination of edges, an expression of the form $c_1 a + c_2 b + c_3 c$, and see when the boundary is zero, i.e. when is

$$\partial(c_1 a + c_2 b + c_3 c) = 0.$$

We can expand the left hand side based on what we said before:

$$c_1 \partial(a) + c_2 \partial(b) + c_3 \partial(c) = c_1(y - x) + c_2(y - x) + c_3(y - x)$$

and grouping together the y and x terms gives

$$(c_1 + c_2 + c_3)y - (c_1 + c_2 + c_3)x = 0.$$

For this to be satisfied the coefficients of y and x must be zero, and that means we require

$$c_1 + c_2 + c_3 = 0,$$

one equation for three unknowns. If we let c_1 be α and c_3 be β, this implies c_2 is $-\alpha - \beta$ and so the set of all loops is $\alpha a - (\alpha + \beta)b + \beta c$, or we can write it as

$$\alpha(a - b) + \beta(c - b).$$

In other words, there are two basic loops, $a - b$ and $c - b$, which I will label z_1 and z_2, that form a basis for the set of all loops (i.e. any other loop in $\mathcal{M}$ is some combination of z_1 and z_2) which I will write as $\{z_1, z_2\}$, and since there are two of them the set of loops is 2-dimensional (this is the 'cyclomatic number' we saw in Chapter 8).

But these two basis loops are not the same in a crucial way: one of them, $c - b$, is the boundary of a face. As such, this loop can be shrunk to a point; *it does not identify a hole*. The other loop however cannot be shrunk to a point, and so it *does* identify a hole (I am slightly mixing the related notions of 'a curve that bounds' and 'a curve that can be shrunk

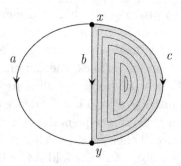

to a point' for the sake of exposition). Let us make this formal: the face A has a boundary, $\partial(A)$, which is a loop, and therefore $\partial(A)$ can be written as a combination of z_1 and z_2. To account for the fact that this loop can be shrunk to a point we set the boundary to this face equal to zero, in other words

$$\partial(A) = c - b = z_2 = 0 \qquad \text{or} \qquad \text{'set } z_2 \text{ equal to zero'.}$$

So our original basis for the set of loops, $\{z_1, z_2\}$, has been reduced to just one non-zero loop, $\{z_1\}$. The size of this reduced set is the first Betti number, denoted b_1, and we have found

$$b_1(\mathcal{M}) = 1.$$

This is often thought of as 'the number of 1-dimensional holes' (which is a little misleading in that the hole itself is not 1-dimensional; rather it is the boundary to the hole that is 1-dimensional). As such, $\mathcal{M}$ has one 1-dimensional hole (as might seem intuitively obvious to you).

The primary motivation of Poincaré's *Analysis situs* was to extend topology to higher dimensions; in fact he opens [110] with the statement *Nobody doubts nowadays that the geometry of n dimensions is a real object*, and we want to get a sense for how Poincaré extended the description in the previous section to manifolds of arbitrary dimension. We will therefore rephrase our definition of b_1 in a way the can be generalized: we first found a basis for the set of loops and labeled them $\{z_1, z_2, \ldots\}$; then we took each face and set its boundary to zero thus reducing the set $\{z_1, z_2, \ldots\}$ to a certain size; this was the first Betti number b_1.

To make this more general, we begin by observing that the manifolds we would like to study are made up of a finite collection of 'elements' of progressively higher dimensions, starting with 0-d vertices, 1-d edges and so on, but also that the boundary of an n-d element of any manifold $\mathcal{M}$ is made up of $(n-1)$-d elements of $\mathcal{M}$ (the boundary of an edge is a set of vertices, the boundary of a face is a set of edges, and so on). The word 'loop' is failing us so we will instead use the word 'cycle', and just like a loop is a sequence of edges with zero boundary we will say an n-cycle is a sequence of n-d elements of $\mathcal{M}$ with zero boundary (a 'loop' would then be a 1-cycle). Now we can give a definition of the nth Betti number of a manifold $\mathcal{M}$: it is the dimension of the set of n-cycles which are not the boundary of an $(n+1)$-d element of $\mathcal{M}$. This definition also leads to an algorithm:

Step 1: find a basis for the set of all n-cycles (i.e. write the general combination of n-d elements and see when the boundary is zero); call this set $\{z_1, z_2, \ldots\}$.

Step 2: take each $(n+1)$-d element and set its boundary equal to zero; this will reduce the set $\{z_1, z_2, \ldots\}$ to a certain size. This size is b_n.

We will use this algorithm to compute the Betti numbers for the sphere and the torus, with two caveats: firstly I am cheating a little in how I represent these surfaces so as to make the presentation clearer (a more legitimate representation is in Figure 15.6 or the next chapter); and secondly the computation of the 0th Betti number, b_0, goes through a little differently than the others and muddies the waters a bit (vertices don't have boundaries as such), so I will just say that the way to interpret b_0 is as the 'number of components of $\mathcal{M}$' [7], which for us is just 1 (e.g. a sphere has only one component or piece).

Let us begin with the sphere $\mathbb{S}^2$, which we have divided into a set of elements of various dimensions and we have given each an orientation (see Figure 15.2).

$$\text{------} \quad b_1 \quad \text{------}$$

Step 1: To find the set of 1-cycles we take the set of 1-d elements, i.e. the edges a and b, form the general combination, $c_1 a + c_2 b$, and set the boundary to zero:

$$\partial(c_1 a + c_2 b) = 0.$$

Since $\partial(a) = y - x$ and $\partial(b) = x - y$ this means we want to solve

$$c_1(y - x) + c_2(x - y) = 0$$

which is the same as $(c_2 - c_1)x + (c_1 - c_2)y = 0$, and so there is only one

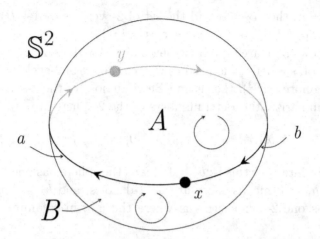

Figure 15.2: Computing the homology of the sphere, $\mathbb{S}^2$.

equation c_1 and c_2 must satisfy:

$$c_2 - c_1 = 0.$$

If we let $c_1 = \alpha$ this means $c_2 = \alpha$ too and so the set of all 1-cycles is $\alpha(a + b)$; we label $a + b$ as z_1, and thus the basis for all 1-cycles is $\{z_1\}$.
Step 2: we take each 2-d element, A and B, and set their boundaries to zero:

$$\partial(A) = 0 \qquad \text{and} \qquad \partial(B) = 0.$$

But $\partial(A) = a + b = z_1$ which means $z_1 = 0$, and similarly $\partial(B) = -a - b = -z_1$ which means the same; there are no non-zero 1-cycles left and hence $b_1(\mathbb{S}^2)$, the first Betti number of the 2-sphere, is 0.

——— b_2 ———

Step 1: to find the set of 2-cycles we take the general combination of 2-d elements, $c_1 A + c_2 B$, and set the boundary equal to zero:

$$\partial(c_1 A + c_2 B) = 0.$$

We know $\partial(A) = a + b$ and $\partial(B) = -a - b$ and this leads to

$$(c_1 - c_2)a + (c_1 - c_2)b = 0$$

and there is only one equation for c_1 and c_2 to satisfy: $c_1 - c_2 = 0$. If we let $c_1 = \alpha$, then $c_2 = \alpha$ and the set of 2-cycles is $\alpha(A + B)$; we label $A + B$ as z_1 and the basis for 2-cycles is $\{z_1\}$.

Step 2: now there are no 3-d elements of $\mathbb{S}^2$ for these 2-cycles to be the boundary of, so there is no need to reduce the set of 2-cycles which stays $\{z_1\}$. Therefore $b_2(\mathbb{S}^2)$, the second Betti number of the 2-sphere, is 1.

To summarize, the Betti numbers of the 2-sphere are

$$b_0(\mathbb{S}^2) = 1, \quad b_1(\mathbb{S}^2) = 0, \quad b_2(\mathbb{S}^2) = 1. \tag{15.1}$$

The way to interpret this is: $b_0 = 1$ means the sphere has one component or piece, $b_1 = 0$ means there are no 1-d holes, and $b_2 = 1$ means the sphere has one 2-d hole: this is where the jam in a jammy doughnut goes.

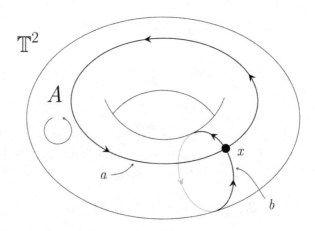

Figure 15.3: Computing the homology of the torus, $\mathbb{T}^2$.

Let us do the same computation for the torus, denoted $\mathbb{T}^2$, which we show in Figure 15.3 with a division into elements, each of which has been given an orientation.

——— b_1 ———

Step 1: There are two 1-d elements, a and b, so to find the set of 1-cycles we set

$$\partial(c_1 a + c_2 b) = 0.$$

But $\partial(a) = x - x = 0$ and the same for $\partial(b)$, so there is no equation for c_1 and c_2 to solve, which means they can be anything we like. If we let

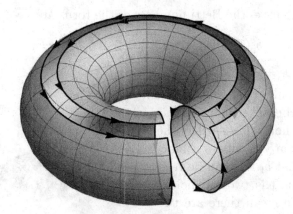

Figure 15.4: The torus $\mathbb{T}^2$ cut along the two basis 1-cycles.

c_1 be α and c_2 be β then the set of all 1-cycles is $\alpha a + \beta b$; we label a as z_1 and b as z_2 and a basis for the set of 1-cycles is $\{z_1, z_2\}$.

Step 2: Now when we look for the 2-d elements, there is only one, A. It is often hard to see that from Figure 15.3, so I have drawn in Figure 15.4 the torus after being cut along the two basis 1-cycles (actually we will see a lot of this sort of thing in the next chapter). If we set its boundary to zero:

$$\partial(A) = 0,$$

then from Figure 15.4 we can see that $\partial(A) = -a - b + a + b$ which is automatically zero, and so there is no reduction on the set $\{z_1, z_2\}$. This means $b_1(\mathbb{T}^2) = 2$.

—————— b_2 ——————

Step 1: There is only one 2-d element, A. The general combination of 2-d elements is just $c_1 A$, and setting its boundary to zero,

$$\partial(c_1 A) = 0$$

gives $c_1 \partial(A) = 0$ but we know $\partial(A)$ is automatically zero so there is no equation for c_1 to solve, and hence c_1 can be anything. If we let $c_1 = \alpha$, the set of all 2-cycles is αA, and letting $A = z_1$ this means a basis for the set of 2-cycles is $\{z_1\}$.

Step 2: There are no 3-d elements of the torus and so there is no reason to reduce $\{z_1\}$; therefore $b_2(\mathbb{T}^2) = 1$.

To summarize, the Betti numbers of the torus are

$$b_0(\mathbb{T}^2) = 1, \quad b_1(\mathbb{T}^2) = 2, \quad b_2(\mathbb{T}^2) = 1. \tag{15.2}$$

As before, $b_0 = 1$ because the torus has only one component, and $b_2 = 1$ because the torus has a single 2-d hole (this is where the air goes when you inflate a bicycle tire). That $b_1 = 2$ can be read as 'the torus has two 1-d holes', by which we mean there are two 1-cycles which detect holes; in the image on the right we see two 1-cycles and the discs they bound. You might say 'but those discs are not part of the torus!' and that is the very point: a hole is an *absence* of something.

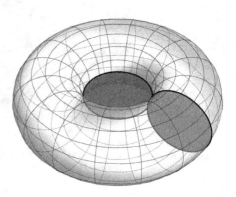

Now we have drunk the fine wine, it is time to reach for the 10-year-old triple-distilled whiskey my Granddad kept for weddings and wakes. Poincaré extended the Euler characteristic to manifolds of arbitrary dimension by counting the number of elements of different dimensions; so if α_m is the number of m-d elements, then

$$\chi(\mathcal{M}) = \alpha_0 - \alpha_1 + \alpha_2 - \alpha_3 + \alpha_4 - \dots$$

We sometimes refer to this as the *Euler-Poincaré characteristic*, and Poincaré showed it is independent of the division used: he argued that just like when we add a vertex to an edge that means the number of vertices and the number of edges both go up by one (so χ remains the same), in the same way if you add an m-d element this divides an $(m+1)$-d element in two, so the alternating sum for χ does not change.

Then he showed that the Euler characteristic can also be written as the alternating sum of Betti numbers:

$$\chi(\mathcal{M}) = b_0 - b_1 + b_2 - b_3 + b_4 - \dots$$

(I encourage you to check this for the sphere and torus, (15.1) and (15.2)). Now not only is this alternating sum an invariant, but also *each Betti number* is an invariant. Not only that, he showed that for surfaces

228 ANALYSIS SITUS.

Donc

$$P_p = P_{h-p}.$$

Par conséquent, pour une variété fermée, les nombres de Betti également distants des extrêmes sont égaux.

Ce théorème n'a, je crois, jamais été énoncé; il était cependant connu de plusieurs personnes qui en ont même fait des applications.

Figure 15.5: Excerpt from Poincaré's *Analysis situs* [95] where he gives his 'duality theorem': *the Betti numbers equidistant from the ends are equal.*

like the sphere and torus and more generally for n-d manifolds that are 'closed' (i.e. finite and without boundary), then the list of Betti numbers is palindromic (Figure 15.5):

$$b_0 = b_n, \qquad b_1 = b_{n-1}, \quad b_2 = b_{n-2}, \ \ldots$$

Since a closed manifold of n dimensions with only one component has $b_0 = 1$, this means we must also have $b_n = 1$; in other words, every such manifold must contain a single n-d hole. But what's more, if the manifold has odd dimension, then the Euler characteristic must be zero! For example a 5-d manifold has

$$\chi = b_0 - b_1 + b_2 - b_3 + b_4 - b_5$$

but if $b_0 = b_5$, $b_1 = b_4$ and $b_2 = b_3$, then everything cancels and $\chi = 0$. We saw this in the last chapter for spheres of odd dimension, and now we see it holds for closed manifolds in general. This is a bad thing by the way: it means we can use the Euler characteristic to distinguish things like spheres and tori in 2-dimensions, but not in 3-dimensions, since everything has the same value for χ. Perhaps we can use the Betti numbers instead? For example, if two manifolds have the same Betti numbers, are they 'the same'?

Poincaré is often described as 'a conqueror, not a coloniser': he would come upon a new topic, transform it completely, and then move on to the next thing, leaving it to his colleagues to iron out the details, but topology was perhaps different. He wrote five supplements [110] to his *Analysis situs* over the next 10 years in response to criticism from some of his peers; in particular Heegaard gave an example of a manifold for which Poincaré's duality theorem of Figure 15.5 failed. Poincaré had overlooked the possibility that manifolds may be 'non-orientable' (they don't have a well defined inside and outside, as we will see in the next chapter), and he needed to clarify several of his arguments, redefine Betti numbers, be more precise about what was included under the term 'manifold'; in fact he didn't actually prove that the Betti numbers are topological invariants. To make homology rigorous, to create from it a systematized and powerful branch of mathematics, we need *eine kleine gruppentheorie*.

In 1926 Pavel Aleksandrov [68] was giving lectures on topology at Göttingen; sitting in the room was Emmy Noether. Aleksandrov said *instead of the usual definition of Betti numbers... she suggested immediately defining the Betti group as the [quotient] group of the group of all cycles by the subgroup of cycles homologous to zero.* This inspired comment lead, through the efforts of Hopf and Veitoris, to the introduction of *homology groups*, which we will try to get a sense of in a manner appropriate for the level of this text (Giblin [49] is good, see also [7]). Nonetheless you might find the last part of this chapter a bit formal; see what you think of it. First we need to clarify exactly what it is we are studying.

Poincaré defined his manifolds as the solutions to equations or inequalities, but the need for precision led him to reformulate everything in terms of what we now call *simplices* (plural of *simplex*), which are like triangles generalized to arbitrary dimensions, an example is in Figure 15.6. We call an n-dimensional simplex an 'n-simplex', and the pattern goes like this: an edge (or '1-simplex') has two vertices as its boundary; a triangle (or '2-simplex') has three edges as its boundary; a tetrahedron (or '3-simplex') has four triangles as its boundary; a 4-simplex will have five 3-simplices as its boundary, and so on (what do you think an 8-dimensional simplex will have as its boundary?). It took some time for the details to be worked out, such as what manifolds can indeed be represented by a collection of simplices (a 'complex'), and it is perhaps ironic that mathematicians felt it necessary to go back to the beginning of topology, to Euler's observations on the Platonic solids from Chapter 13, in order to make progress; for example the representation of the sphere in terms of simplices is the elemental tetrahedron (Figure 15.6).

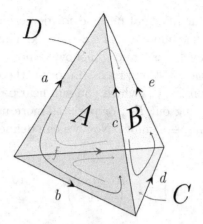

Figure 15.6: A tetrahedral model of the sphere in terms of simplices.

Let's look at the edges, or 1-simplices, first. Previously we talked about the 'general combination' of 1-simplices, which is the expression

$$c_1 a + c_2 b + c_3 c + c_4 d + c_5 e + c_6 f$$

where c_1, c_2 and so on are integers. Expressions of this type are called '1-chains', and are formal in the sense that they don't need to make particular sense with respect to the tetrahedron: for example $c - a + f$ and $d + e - c - b$ can be traced in Figure 15.6, but something like $b - 2f + 3e$ cannot; nonetheless it is a 1-chain. We can add two 1-chains and get another 1-chain ($a + b - 2e$ added to $e - a + c$ gives $b + c - e$ for example), there is an 'identity' 1-chain (which we just label 0), and every 1-chain has an inverse (if you add $a - b - 2c$ and $-a + b + 2c$ you get 0). As such, the set of 1-chains forms a *group*, as we saw in Chapter 7; we label it $\mathcal{C}_1$.

This group contains a subset, the set of all 1-cycles, which are just 1-chains with zero boundary. If you add two 1-cycles you get another 1-cycle (for example in Figure 15.6 adding $b + c - a$ and $d + e - c$ gives $b + d + e - a$); 0 is a 1-cycle (it is a 1-chain with zero boundary); and every 1-cycle has an inverse. As such, the set of all 1-cycles is also a group, and is therefore a subgroup of $\mathcal{C}_1$; we label it $\mathcal{Z}_1$.

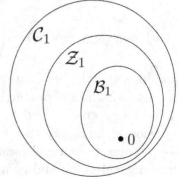

Finally $\mathcal{Z}_1$ contains a subset: the set of all 1-cycles which are the boundary of a 2-simplex. Again this set of all boundaries is a subgroup of $\mathcal{Z}_1$; we label it $\mathcal{B}_1$.

Think back to our informal method for detecting a hole: we seek a 1-cycle which is not the boundary of a 2-simplex. This means we want to take the group of 1-cycles, $\mathcal{Z}_1$, and somehow 'remove' the elements of $\mathcal{B}_1$. The proper description of this process in group theory is the 'quotient' of $\mathcal{Z}_1$ and $\mathcal{B}_1$, written $\mathcal{Z}_1/\mathcal{B}_1$, which you will hear described as 'dividing out by $\mathcal{B}_1$' or 'factoring out $\mathcal{B}_1$', or perhaps more usefully we 'send $\mathcal{B}_1$ to zero'; I will use the '$\div$' symbol. Now we can define the first homology group, $\mathcal{H}_1$, as

$$\Big(\text{First homology group}\Big) = \left(\begin{array}{c}\text{Group of}\\ \text{1-cycles}\end{array}\right) \div \left(\begin{array}{c}\text{Group of boundaries}\\ \text{of 2-simplices}\end{array}\right),$$

or 'cycles divided by boundaries'. This group will typically (!) be several copies of the group of integers with addition, $(\mathbb{Z}, +)$, just like how the fundamental group of the torus from Chapter 7 was two copies of $(\mathbb{Z}, +)$ (in fact *very* like, we are abel to show). The number of copies of $(\mathbb{Z}, +)$ is the first Betti number, b_1.

Now we have the necessary terminology and notation to generalize: $\mathcal{C}_n$ is the group of all formal combinations of n-simplices, and the dimensional hierarchy is usually written like this:

$$\cdots \xrightarrow{\partial} \mathcal{C}_3 \xrightarrow{\partial} \mathcal{C}_2 \xrightarrow{\partial} \mathcal{C}_1 \xrightarrow{\partial} \mathcal{C}_0,$$

the '$\xrightarrow{\partial}$' symbol signifying that the boundary of an element of one group is an element of the group to the right. This boundary will be a cycle; in fact each $\mathcal{C}_n$ contains a subgroup $\mathcal{Z}_n$ (the group of n-cycles) which in turn contains a subgroup $\mathcal{B}_n$ (the n-cycles which are boundaries of an $(n+1)$-chain). The nth homology group is then

$$\Big(n\text{th homology group}\Big) = \left(\begin{array}{c}\text{Group of}\\ n\text{-cycles}\end{array}\right) \div \left(\begin{array}{c}\text{Group of boundaries}\\ \text{of }(n+1)\text{-simplices}\end{array}\right)$$

and its size is the nth Betti number, b_n.

The reason to phrase everything in terms of groups is that we know *a lot* about groups, and in fact it was via this group theoretic expression of homology that the Betti numbers (and homology groups) were proven to be topological invariants. Poincaré's introduction of the fundamental group and homology, and the subsequent development of homology groups, was the beginning of Algebraic Topology, which has been one of the major developments of 20th-century mathematics and arguably has become the modern expression of Topology as a whole.

CHAPTER 16

Classification

W HAT MANIFOLD IS THIS? By which I mean: if presented with a new manifold, what is it 'the same' as? How can we recognize it in terms of familiar or standard objects like spheres and cylinders? We want to be able to classify *all* the manifolds of a certain class, so when we meet a new member of that class we know which box to put it in. The first great success in this direction was classifying all surfaces through a progression of gradually stronger results in the second half of the 19th century [96]; this might seem like a modest achievement, but surfaces can be weird and wonderful, like the examples below which look like they have come from a medieval bestiary:

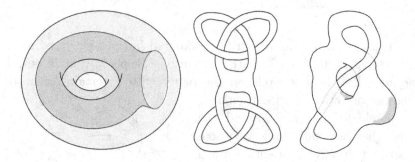

Figure 16.1: What surfaces are these?

What's more, some surfaces don't have an inside and an outside, as we will see, which makes visualizing them difficult. We will describe a method of classifying surfaces using the 'word' representation, for which we first need to introduce some ideas like orientability and connected sums, but then we will move beyond surfaces to consider classifying

DOI: 10.1201/9781003455592-16

manifolds of higher dimensions which then leads to one of the most famous problems of 20th century mathematics: the Poincaré conjecture.

Consider a rectangle where we have 'identified' the two short ends, by which I mean we consider pairs of points on those ends to be the same point, according to the arrows shown. If the rectangle were made of paper, I could glue the two short ends together so the arrows line up, and get a (portion of a) cylinder.

Suppose we instead identify the ends of a rectangle but with the two arrows pointing in opposite directions; if we join the ends of a piece of paper so the arrows lined up, we would need to put a 'twist' in it.

This is known as a 'Möbius strip', discovered independently by Möbius and Listing in 1858 [109], with the curious property that it only has *one side*. If I draw a circle on the paper with a certain orientation, clockwise or anti-clockwise, then drag that circle all the way around the strip, when we come back to where we started the circle's orientation will have reversed. For this reason we say the Möbius strip is *non-orientable*, whereas a cylinder or sphere is *orientable*.

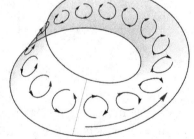

Is the cylinder the same as the Möbius strip? If we look at the cylinder we see it has two edges, or 'boundary components', whereas the Möbius strip has only one. If two manifolds are to be the same as each other then surely they would need to have the same number of boundary components; that number is an *invariant* (it doesn't change as we continuously deform the manifold).

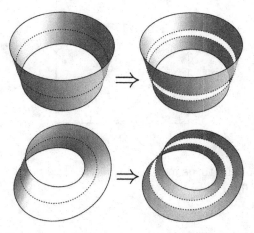

There is another interesting difference between the cylinder and Möbius strip: if you took a scissors and cut along the central dashed line of the cylinder, then it would separate into two pieces; two cylinders to be precise. On the other hand if you cut along the dashed line of a Möbius strip it does not separate into two, it is still a single piece of paper; what is it? Another Möbius strip? Well it now has two boundary components, so it is a cylinder.

In other words, when presented with a new object (such as a Möbius strip cut in 'half'), we can identify it by studying its invariants. This is the essence of 'classification', and Möbius also gave us, in 1863, the first classification theorem for surfaces but funnily enough only those that are orientable. He showed that all orientable surfaces are the same as the 'connected sum' of simple surfaces like spheres and tori: you take two surfaces, remove a disc from each, add a little tube, then smooth out the join. On the right we show the connected sum of two tori, called a 2-torus, and in fact all the surfaces in Figure 16.1 are topologically the same as a 2-torus. In 1888 von Dyck classified all surfaces, both orientable and non-orientable, in terms of connected sums however the first proof considered rigorous and complete by modern standards is due to Dehn and Heegaard in 1907 [96]. That said, the clearest way to understand how someone can possibly classify all surfaces is due to Brahana

from 1921 and that is the method we will describe here (Carlson [21] is particularly good for this, as is [42]).

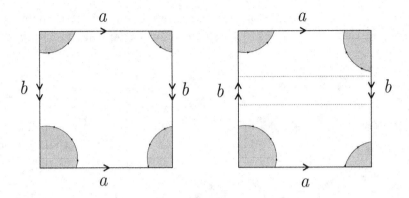

Figure 16.2: The plane models of the torus and Klein bottle. The neighborhood of a point is shown in light gray.

First we must clarify exactly what we mean by 'surface'; just saying 'a 2-dimensional object' would open the door to some of the fractal beasts we saw in Chapter 4 and we'd never make any progress. Instead we will say: a surface is any finite manifold where the neighborhood of a point (which just means the little part of the manifold around that point) is topologically the same as a disc. Thus a 2-torus is a surface, as is a sphere, but a Möbius strip is not a surface since it has a boundary (a point on the boundary will not have a neighborhood that is the same as a disc; we will come back to surfaces with boundary later).

Drawing surfaces can get pretty intricate so instead we use the notion of 'plane model' introduced by Klein in the 1880s: we take some polygon, like a square, and identify the edges in pairs using arrows to specify which points are identified with which. Two key examples are in Figure 16.2: if you take the square on the left and imagine it made of stretchy rubber, then we could join the two sides that have a single arrow to make a cylinder (like we did before), and then join the two ends with double arrows to make a torus; in fact we have already seen this in Figure 15.4.

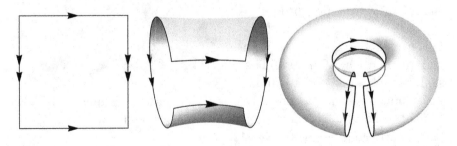

On the other hand, if the square on the right of Figure 16.2 were made of stretchy rubber then no matter how stretchy it was you would not be able to join the edges according to the arrows; nonetheless this is still a surface according to our definition since the neighborhood of any point is a disc. The reason we can't join the edges is that this plane model contains a Möbius strip (shown in dashed lines), and so this surface is not orientable. It is known as the 'Klein bottle' (you *can* join the edges if you let the Klein bottle sit in four-dimensional space, but not three; if you have seen pictures of weird tubes that pass through themselves they are not the Klein bottle, they are its shadow). It is easy to calculate the Euler characteristic from the plane model; for example both the torus and Klein bottle have $\chi = 0$ (which shows that the Euler characteristic alone is not enough to distinguish these two surfaces).

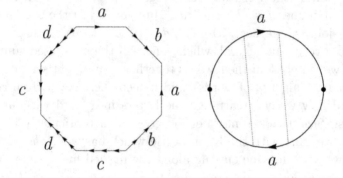

Figure 16.3: The plane models of the 2-torus and projective plane.

You won't be surprised to hear plane models can be very elaborate, let me present two more examples in Figure 16.3. On the left is an octagon with four pairs of edges; if you were able to stretch the octagon enough you could join the edges in 3-d space according to the arrows to make a 2-torus. On the other hand, the plane model on the right cannot be joined up in 3-d space according to the arrows no matter how much you stretch it; like the Klein bottle this surface contains a Möbius strip (for example if we cut out the pair of dashed lines), and so it is not orientable. This surface is called the 'projective plane'. We essentially have all we need now: *every surface is the connected sum of either tori (if orientable) or projective planes (if not)* (by an abuse of language we include the sphere here as 'the connected sum of zero tori'). This is the famous 'classification' theorem for surfaces, and it is pretty strong: no matter how weird or bizarre your surface is, it must be the same as some number of tori or projective planes joined together, that's it. How can such a thing be proved?

The big idea in Brahana's proof is hinted at in the diagrams of the plane models: see how we have labeled the edges with a letter, using the same letter on the edges that are to be joined. If we start at some edge and work our way around clockwise, writing down the label of each edge with the convention 'x' if the arrows on that edge are pointing clockwise and 'x^{-1}' (we say 'x inverse') if the arrows are pointing anti-clockwise, we have the 'word' representation of the surface. For example, from Figure 16.2 the word for a torus is $aba^{-1}b^{-1}$ and the word for the Klein bottle is $aba^{-1}b$; from Figure 16.3 the word for the projective plane is aa. The actual labels themselves don't particularly matter, the word $efe^{-1}f^{-1}$ or $d_1d_2d_1^{-1}d_2^{-1}$ is still a torus.

Now look at the word for the octagonal plane model on the left of Figure 16.3: it goes $aba^{-1}b^{-1}cdc^{-1}d^{-1}$, but this is just the word $aba^{-1}b^{-1}$ (a torus) joined together with $cdc^{-1}d^{-1}$ (also a torus), and the surface itself is a 2-torus. In general, when you form the connected sum of two surfaces we simply join their words together. You might see how we are developing an 'algebra of words', and indeed there are certain rules for when and how we can rearrange the letters in a word without changing the surface it represents. Let me give you a particularly important example: we start with the Klein bottle which has word $aba^{-1}b$ (on the left below), and imagine cutting along the dashed line. We are allowed cut surfaces as long as we remember to glue them back together again; in order to remember what was joined with what we add arrows and the label 'c'.

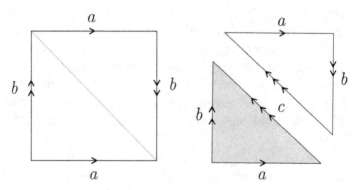

Now taking the shaded triangle we can imagine it made of actual paper and just flip it over vertically, bearing in mind what this does to the arrows; we move it so it lies beside the other triangle.

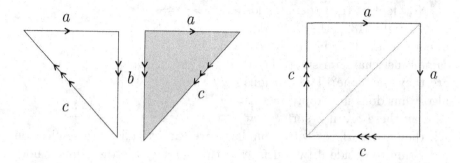

The two sides marked 'b' are side by side, and the arrows match, so we can simply glue them together to get a plane model with 4 sides, only now the word is $aacc$; but aa is a projective plane, and so is cc, so this means the Klein bottle is just the connected sum of two projective planes!

Now we can see a method developing: if you have a word that goes like $x \ldots x \ldots$ then we can use the 'cut and glue' approach from above to rewrite the word as $xx \ldots$ which is the connected sum of a projective plane and something else, but then we look at the 'something else' and apply the same reasoning. In this way we can systematically reorganize the letters of any word until we either have something like $aabbcc \ldots$ or $aba^{-1}b^{-1} \ldots$, which means the surface is the connected sum of projective planes or tori.

What we don't yet know is *how many* tori or projective planes; this is where the Euler characteristic comes in. In the orientable case, the number of tori in the connected sum is called the 'genus', so a 2-torus has genus 2 (or we could say it has two 'holes' in the intuitive sense). The Euler characteristic χ is related to the genus g via $\chi = 2 - 2g$; note that since the sphere has Euler characteristic 2 this includes the sphere as a 'torus with no holes', or genus 0 surface. In the non-orientable case χ also determines the number of projective planes, so finally we have our algorithm for classifying a surface:

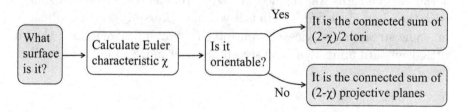

Now look at the two manifolds on the right [132]: they are both orientable and they both have the same Euler characteristic (-1), so are they the same? The previous algorithm does not apply in this case as these are not surfaces at 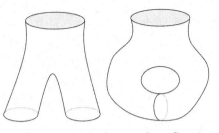 all, they are 'surfaces with boundary'; we can extend our classification algorithm to decide if two surfaces with boundary are the same or not:

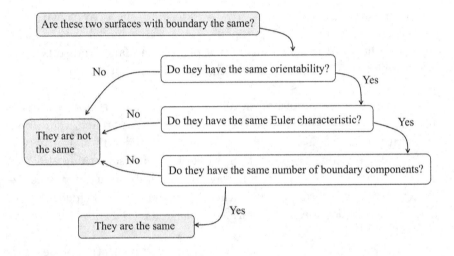

Surfaces: done.

But this is how mathematics works: as soon as one question is answered, many more are asked. If we can classify all surfaces, which are 2-dimensional manifolds, can we classify all 3-dimensional manifolds? How can we even conceive of 3-d manifolds, and why would we want to?

To intuitively grasp a 3-d manifold, let's look again at our work with surfaces. The torus was defined in Figure 16.2 as the inside of a square with the edges identified, and in fact you have probably seen this already: on some streaming services, if you want to search for something then a little keyboard appears on the screen and you move a cursor around.

A	B	C	D	E	F	G	H	I	J	K	L	M
N	O	P	Q	R	S	T	U	V	W	X	Y	Z
1	2	3	4	5	6	7	8	9	0	space	del	

If you are over at 'M' on the right, then pressing right once more will make the cursor appear at 'A' on the left; in the same way if you are at 'D' and press up then you will appear at '4' on the bottom; when you move off one side of the square then you reappear at the opposite side and so this keyboard is on a torus, specifically, a 2-d torus (see [130] for other games we can play on the torus and friends).

Now let's go up a dimension: take the inside of a cube, and iden- tify the opposite faces in the sense that if you were moving along and you reached one face then you would reappear at the opposite face (the two black dots labeled A in the image are identified so they are literally *the same point*). This is the 3-d torus, and if you were standing in the middle of a 3-d torus then you would behold a mesmerizing sight (see Figure

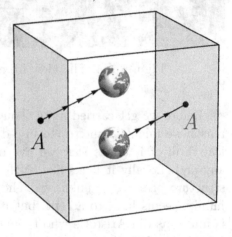

16.4). Imagine shining a torch and a ray of light coming out of your torch, reaching one face of the cube and then reappearing on the face behind you; if you looked straight ahead at one face of the cube, you would see the back of your own head. In fact you would see yourself re- peated infinitely often in all directions, and every copy of yourself would have their back turned to you; if you tried to move forward to meet your copy, then they would all move away from you, infinitely many of them.

In fact this is just one possibility for a 3-d manifold (see the excel- lent Weeks [130] or [3, 116]): we could also start with a cube and then identify or 'glue' opposite faces with a twist, or a flip, or even start with another Platonic solid like a dodecahedron and glue opposite faces to- gether, or perhaps take the interior of copies of a surface and glue their boundary together, like we did in Chapter 14 to describe the 3-sphere; the possibilities are endless!

Figure 16.4: The view from inside a 3-d torus [134].

Before we get carried away though, let's remind ourselves that one reason we might be interested in 3-d manifolds, beyond the mathematical thrill of it all, is that we live in a 3-d universe (the spatial part anyway). Locally it looks like Euclidean space, $\mathbb{R}^3$, but the large-scale structure of space is unknown. The natural assumption that space is infinite seems hard to accept, but if you suppose instead the universe is finite (as did Aristotle who thought the universe was the inside of a large ball), does that mean the universe has a boundary? What is outside the ball? Perhaps 'in the large' our universe might instead connect up in a way that makes it finite but without boundary. In Riemann's 1854 *habilitationsschrift* he suggested our universe might be a 3-sphere, and Schwarzschild in 1900 considered the universe as a 3-d torus. When Einstein introduced the notion of spacetime curvature with his General Relativity this led to cosmological models of the universe [34] as either positively curved, negatively curved, or flat, only one of which is finite and without boundary, but this is too limited and some cosmologists are again returning to the idea that the universe may have a more complicated global topology such as a 3-d torus, even looking for patterns in the distribution of galaxies as evidence [130]. Perhaps if we look deep enough into space we will eventually see our own little blue watery home, floating all alone in the cosmos.

To classify all 3-d manifolds is too big an ask, so instead let's start with something more modest: if we are presented with a mystery 3-d manifold that is closed (i.e. finite and without boundary), can we decide if it is a 3-sphere or not? We have several invariants at our disposal now, some are numbers (like the Euler characteristic or Betti numbers), some are properties (like orientability or connectedness), and some are structures (like the fundamental group), but the Euler characteristic is no longer any use, since all 3-d closed manifolds have $\chi = 0$, as we saw in the last chapter. Instead Poincaré, in his *Analysis situs*, thought his new technique of homology would be enough: if a closed 3-d manifold has the same Betti numbers as a 3-sphere, then it *is* a 3-sphere. However in his 5th supplement, written in 1904 [110], he constructed an elaborate 3-d manifold that had the same Betti numbers as a 3-sphere, but different fundamental group; therefore it is not a 3-sphere.

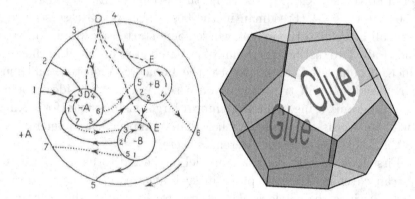

Figure 16.5: Two different representations of Poincaré's *homology sphere*.

His construction was quite complicated, as you can see from his diagram in Figure 16.5 on the left (which reminds me of the spaghetti meatballs I had for dinner the other day): the discs are supposed joined by cylinders, like the connected sum we described previously, to form a surface (a 2-torus in fact); the interior of that surface is a 3-d manifold with boundary; then you take two copies of this manifold and glue them along their boundary [102]. This is called the *homology sphere*, and fortunately for us it was subsequently shown to be equivalent to the manifold you get if you take the interior of a dodecahedron and glue opposite faces in the manner shown in Figure 16.5 on the right.

Poincaré turned instead to his fundamental group, and revised his question: if a closed 3-d manifold has the same fundamental group as the

3-sphere, can we conclude it *is* a 3-sphere? The fundamental group of the 3-sphere is trivial: all closed loops can be shrunk to a point, and today we call a manifold with this property 'simply connected'. Poincaré's question is then: if a closed 3-d manifold is simply connected, must it be a 3-sphere?

This is the last question he poses after 250 or so pages of *Analysis situs* and its five supplements; he does not pursue it, saying *this question would carry us too far away*. He should have said: Statement A.

Poincaré's influence was immense, so naturally everyone wanted to prove the Poincaré conjecture, as his question became known. It was extended to arbitrary dimension ('if a closed n-d manifold is simply connected then it must be an n-sphere') and curiously the higher dimensions were resolved soonest [82]: the conjecture is true for $n \geq 5$ (Smale in the 1960s) and $n = 4$ (Freedman in the 80s). The $n = 3$ case seemed to resist all attack, to the extent where some started to wonder if it was true at all. Two distinct programs of research were initiated in the 1980s which, if completed, would resolve the Poincaré conjecture and much more besides; I will try to give you a sense of the 'big ideas' behind these programs using as little technical jargon as possible (for example I will use words like 'finite' rather than 'compact'), and refer you to [130, 116, 110, 82] and the references therein.

First look back at the plane model of the 2-d torus in Figure 16.2. We can cover or 'tile' the plane in such squares without any gaps or overlaps; since the plane is flat we say we can 'give' the torus a flat geometry (incidentally since the plane model for the Klein bottle is also a square we can give it a flat geometry too). The 2-torus on the other hand has an octagon as its plane model, Figure 16.3, and we can't tile the plane with octagons. What we can instead do is tile the *hyperbolic* plane with octagons, just like we described in Chapter 2 (see Figure 2.2). As such we say we can give the 2-torus a hyperbolic geometry. In fact any surface can be given the geometry of the 2-sphere, the (flat) Euclidean plane, or the hyperbolic plane.

Now look at the cube model of the 3-d torus. We can fill 3-d Euclidean space with such cubes without gaps or overlaps (as Plato knew), and so we say we can 'give' the 3-d torus a Euclidean geometry. On the other hand the dodecahedron model of Poincaré's homology sphere, Figure 16.5, cannot fill Euclidean space but it can fill the 3-sphere, and so we say we can give it a spherical geometry. But if we started with the dodecahedron and identified the faces slightly differently, Figure 16.6 on the left, then we would need to put this in 3-dimensional hyperbolic space (see Figure 2.3) for copies of it to fit together nicely; we can give this 3-d manifold a hyperbolic geometry. This idea of 'giving a geometry' to a manifold is called *geometrization*, only now in 3-d we need more than spherical, Euclidean and hyperbolic: there are in fact eight different geometries needed to describe all closed 3-d manifolds. This is the content of Thurston's *geometrization conjecture* (1982). An example of a new geometry is in Figure 16.6 on the right: we take the 3-d region between two concentric spheres and identify their surfaces. If Thurston's geometrization conjecture were true, then the Poincaré conjecture would also be true: loosely, some geometries are simply connected but not finite (such as Euclidean space), some are finite but not simply connected (like the concentric sphere model below), but only one is finite *and* simply connected: the 3-sphere.

Figure 16.6: These 3-d manifolds can be given one of Thurston's eight geometries (compare with Figures 16.5 and 14.13).

For the second 'big idea' needed to prove Poincaré's conjecture, and the last big idea in this book, it is fitting we draw things to a close with the silver thread that has run throughout: curves in the plane. Consider the curve on the left of Figure 16.7; at each point there is a curvature, as described in Chapter 3, and we draw the normal vector at every point scaled by the curvature. We now imagine we 'flow' the curve according to its curvature, by which I mean we move the curve according to these vectors to get a new curve, and then repeat the process again and again to give a sequence of curves, on the right of Figure 16.7. The thing to observe is that the curves are gradually evolving to a circle; the curvature flow is 'spreading the curvature out', and leading to a curve with constant curvature (i.e. a circle). This curvature flow is not changing the topology of the curve, it is still connected, but its geometry is becoming simpler.

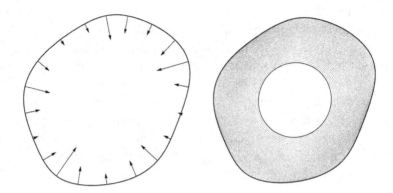

Figure 16.7: The curvature flow of a curve in the plane.

We can do the same thing with surfaces (ever wonder why pebbles on the beach are smooth and round?) and for manifolds, only now there are different types of curvature to choose from, like the Gauss curvature we discussed in Chapter 3 or the Germain/mean curvature from Chapter 12. In 1982, Hamilton [54] introduced a flow of 3-d manifolds due to the *Ricci* curvature (which we will denote R, see Figure 16.8), the very same Ricci curvature that finds its way into Einstein's field equations of General Relativity from Chapters 10 and 12. Under this 'Ricci flow' the curvature spreads out like above, and so R is changing in time but also from point to point; the equation for the flow is

$$\left(\begin{array}{c} \text{Rate of change of } R \\ \text{with respect to time} \end{array} \right) = \left(\begin{array}{c} \text{Second derivatives of } R \\ \text{with respect to space} \end{array} \right) + \left(\begin{array}{c} \text{Other} \\ \text{terms} \end{array} \right)$$

and this is a partial differential equation, in fact it is essentially the heat or diffusion equation from Chapter 12 (12.4). As we 'melt' the manifold, the curvature spreads out in a way that makes the geometry simpler but keeps the topology the same, hopefully settling onto one of Thurston's eight geometries. If whenever you melt a simply-connected manifold you end up with a (geometrical) 3-sphere, then what you started with must have been a (topological) 3-sphere! But there's many a slip 'twixt cup and lip: if the curvature is negative, then as the manifolds flow they

7.5 Corollary. *The scalar curvature R satisfies the evolution equation*

$$\frac{\partial}{\partial t} R = \Delta R + 2 g^{ij} g^{kl} R_{ik} R_{jl}.$$

7.6 Corollary. *If the scalar curvature R > 0 at t = 0, then it remains so.*

Proof. The term $g^{ij} g^{kl} R_{ik} R_{jl}$ is just the norm squared of the Ricci curvature, and hence is always positive. The result now follows from the maximum principle for the heat equation. This simple example is a model for our subsequent a priori estimates. It also shows why the evolution equation "prefers" positive curvature.

Figure 16.8: Excerpt from Hamilton's *Three-manifolds with positive Ricci curvature* [54]. The symbol ΔR is short for the Laplacian of R (12.4), and g is the metric (Chapter 3 and 12).

can develop singularities, pinch points, and then the flow breaks down; resolving these technical issues seemed impossibly complex.

Grigori Perelman was a bright star of the Russian specialist mathematical schools. Though the Soviet Russia he grew up in was riddled with anti-Semitism (for example Leningrad University's Mathematics Department had a yearly quota of only two Jewish students [48]), Perelman's talent could not be denied after he achieved a perfect score at the 1982 International Mathematical Olympiad. His early career showed great promise, but by the late 1990s he seemed to have dropped off the map; certainly no-one thought he was working on the Poincaré conjecture. Then in 2002/2003, to everyone's astonishment, Perelman published on the arXiv three papers in which he completed Hamilton's program by performing 'surgery' on manifolds: if a singularity develops we stop the flow, chop off the troublesome part and replace it with something nicer, then start the flow again (for example his second paper is titled *Ricci flow with surgery on three-manifolds* [92]). Perelman's work is a *tour de force* of analysis, proving that Thurston's geometrization conjecture is true and as a consequence the Poincaré conjecture is also true (although we should perhaps now call it the Perelman theorem or Perelman-Hamilton). This shining accomplishment was woven out of beautiful and complex ideas from the Four Corners of Mathematics: curvature (Geometry), the fundamental group (Algebra), partial differential equations (Calculus), and simply connected manifolds (Topology). Remove one of those threads and everything unravels.

As well as other prizes, Perelman turned down the Fields medal, the highest honour in mathematics, saying that Hamilton deserved as much credit as he did. We might go further: credit should also go to Poincaré and Noether, to Gauss and Pythagoras, and all the men and women over the centuries who have fallen for the profound beauty that is the finest creation of the human mind: Mathematics.

Conclusion

MATHEMATICS didn't begin with Pythagoras, and it certainly didn't end with Perelman. Mathematics has been around for a long time and it isn't going anywhere; if anything, as we enter the age of artificial intelligence and machine learning the field is more vital and vibrant now than it ever has been. What is changing perhaps is the

face of our subject, with more female mathematicians receiving opportunity and recognition. For example the Fields medal was awarded to a female mathematician for the first time in 2014, to Maryam Mirzakhani *for her outstanding contributions to the dynamics and geometry of Riemann surfaces and their moduli spaces.* If we look at some of her work, for example [83], we see it is full of terms we recognize from this book: closed curves and geodesics, genus and curvature, groups and hyperbolic geometry, vector spaces and homology. Her work, and the work of all mathematicians, builds on the results and ideas of previous mathematicians, and in turn

Maryam Mirzakhani (1977-2017) [61]

contributes to the results and ideas of future mathematicians. This is how it has always been, right from the beginning, and how it must continue to be; who knows what mathematics is still to come? When Archimedes wrote of his 'method' [41] he spoke to his *successors [who] will be able to discover new theorems...which have not yet occurred to me.* The ghost of Archimedes calls out to us! It is our turn to take our place in the line, shoulder to shoulder with those who have come before and those who are to follow. Mathematics is hard but the rewards are great, and sometimes in life we must grasp the nettle. Even Gauss needed some self-encouragement from time to time; his diary entry for July 9th 1796 reads: *the sum of three squares in continued proportion can never be a prime: a clear new example which seems to agree with this. Be bold!*

Figure C.1: The mathematicians associated with the main developments in each chapter, from Pythagoras to Perelman.

	Geometry				Algebra				Calculus				Topology			
	The Beginnings	Non-Euclidean	Curves & Surfaces	Fractals	The Beginnings	Complex Numbers	Abstract Algebra	Linear Algebra	The Beginnings	Solar System	Maxima Minima	PDE's	The Beginnings	Degree	Homology	Classify
Classical to 500's	Pythagoras Euclid Apollonius Hypatia			Apollonius Pappus	Early numbers; Diophantus				Archimedes	Eudoxus Apollonius						
Medieval	Hindu Islamic		Oresme		Al-Khwārizmī											
1600's	Fermat Descartes		Descartes Huygens Newton	Leibniz	Cardano Viète Descartes Fermat	Bombelli			Fermat Newton Leibniz	Kepler; Newton	Fermat					
1700's			Euler			Euler					Euler Lagrange	D'Alembert Bernoulli	Euler			
1800's		Gauss Bolyai Lob'vsky	Gauss Riemann	Sierpiński		Gauss	Ruffini Abel Galois	Hamilton Grassman Cayley	Cauchy	Laplace Somerville		Laplace Fourier Germain	Gauss Bonnet	Cauchy		Möbius Klein
1900's to present		Poincaré	Einstein	Mandelbrot			Poincaré Noether			Poincaré Einstein	Noether	Einstein	Poincaré	Poincaré Whitney	Poincaré Noether	Poincaré; Thurston Hamilton Perelman

Acknowledgements

My thanks to Professors Brien Nolan, Colin McInnes, Tom Sherry and Andrew Osbaldestin, for seeing some potential in me and giving me the opportunity to become a mathematician.

My thanks to Dr. Graham Elliott for reading this text, and for letting me have his Linear Algebra notes 15 years ago; also thanks to (future Dr.) Dougie Howells for reading this text and making many intelligent suggestions for improvements; and thanks to all my colleagues at the University of Portsmouth.

My thanks to 'the lads' who I have known for many years and many beers.

My thanks to my family and especially my mum and dad, Denise and Dermot, for unwavering belief, support and love.

Finally my biggest thanks and appreciation to my two fantastic daughters, Emily and Rachel, and my wonderful wife, Maria, for bringing true happiness and meaning to my life.

Bibliography

[1] 3Blue1Brown: But what is a Fourier series? https://www.youtube.com/watch?v=r6sGWTCMz2k

[2] A. Sherwood. Regular tiling of the hyperbolic plane by quadrilaterals https://commons.wikimedia.org/wiki/User:Tamfang/H2

[3] C. Adams and R. Franzosa. *Introduction to topology; pure and applied.* Pearson Prentice Hall, 2008.

[4] A. Alexander. *Duel at dawn; heroes, martyrs, and the rise of modern mathematics.* Harvard University Press, 2010.

[5] M. Anthony and M. Harvey. *Linear Algebra: concepts and methods.* Cambridge University Press, 2012.

[6] H. Anton, I.C. Bivens, and S. Davis. *Calculus (Late Transcendentals).* John Wiley and Sons, 2010.

[7] M.A. Armstrong. *Basic Topology.* McGraw-Hill, 1979.

[8] J. Arroyo, O.J. Garay, and J.J. Mencía. When is a periodic function the curvature of a closed plane curve? *The American Mathematical Monthly,* 115(5):405–414, 2008.

[9] B. Rosenfeld. Apollonius Conics online: http://skatok.s3-website-us-east-1.amazonaws.com/Books1-7new.pdf

[10] J. Barrow-Green. *Poincaré and the three body problem.* American Mathematical Society, 1997.

[11] E.T. Bell. *Men of Mathematics: The Lives and Achievements of the Great Mathematicians from Zeno to Poincaré.* Simon and Schuster, 1937.

[12] M.V. Berry. Regular and irregular motion. *AIP Conference Proceedings,* 46:16–120, 1978.

[13] Bill Tavis. `https://www.kickstarter.com/projects/billtavis/mandelmap-poster-a-vintage-style-map-of-the-mandel`

[14] Black Heroes of Mathematics Conference 2023. `https://ima.org.uk/22240/black-heroes-of-mathematics-conference-2023/`

[15] C. Booth. *Hypatia: Mathematician, Philosopher, Myth*. Fonthill, 2016.

[16] C.B. Boyer. *A history of mathematics*. John Wiley and Sons, 1991.

[17] S.L. Brunton and J.N. Kutz. *Data-driven science and engineering; machine learning, dynamical systems, and control*. Cambridge University Press, 2022.

[18] J. Burridge. Spatial evolution of human dialects. *Phys. Rev. X*, (7):031008, 2017.

[19] Edited by J.M. Blackledge, A.K. Evans, and M.J. Turner. *Fractal Geometry: Mathematical Methods, Algorithms, Applications*. Horwood Publishing Ltd., 2002.

[20] (c) mapz.com – Map Data: OpenStreetMap ODbL.

[21] S.C. Carlson. *Topology of surfaces, knots and manifolds: a first undergraduate course*. John Wiley & Sons, 2001.

[22] E. Chladni. *Die akustik*. Leipzig, 1802.

[23] Claire Cock-Starkey. *Hyphens and Hashtags; the stories behind the symbols on our keyboards*. Bodleian Library, 2021.

[24] LIGO/Virgo collaboration. Observation of gravitational waves from a binary black hole merger. *Phys. Rev. Lett.*, (116):061102, 2016.

[25] Creative Commons Attribution 3.0 Unported.

[26] Credit: ESA/Rosetta/NavCam – CC BY-SA IGO 3.0.

[27] David Eck. `https://math.hws.edu/eck/js/mandelbrot/MB.html`

[28] P. de Fermat. *Arithmeticorum libri sex, et de numeris multangulis liber unus. Cum commentariis C.G. Bacheti & observationibus D. P. de Fermat.* Toulouse: Bernard Bosc, 1670.

[29] P. de Fermat. *Varia Opera Mathematica.* 1679.

[30] D.E. Joyce. Euclid's Elements online: `http://aleph0.clarku.edu/~djoyce/java/elements/elements.html`

[31] R. Descartes. *The Geometry, translated by D.E. Smith and M.L. Latham.* New York: Dover, 1954.

[32] F. Diacu and P. Holmes. *Celestial encounters; the origins of chaos and stability.* Princeton Science Library, 1996.

[33] A. Dick. *Emmy Noether, 1882-1935.* Birkhäuser, 1981.

[34] R. d'Inverno. *Introducing Einstein's Relativity.* Clarendon Press, 1992.

[35] M.P. DoCarmo. *Differential Geometry of Curves and Surfaces.* Prentice-Hall, 1976.

[36] A. Einstein. Zur elektrodynamik bewegter körper. *Annalen der Physik,* 4(17):891–921, 1905.

[37] J. Foster et al. Systematically improving espresso: insights from mathematical modeling and experiment. *Matter,* 2(3):631–648, 2020.

[38] L. Euler. *Introductio in analysin infinitorum, Volume 1.* Euler Archive - All Works. 101., 1748.

[39] L. Euler. Recherches sur la courbure des surfaces. *Mém. Acad. Sci.Berlin,* (16):119–143, 1760.

[40] K. Falconer. *Fractal Geometry; Mathematical Foundations and Applications.* John Wiley and Sons, 1990.

[41] J. Fauvel and J. Gray. *The history of mathematics: a reader.* Palgrave MacMillan, 1987.

[42] P.A. Firby and C.F. Gardiner. *Surface Topology.* Woodhead Publishing, 2001.

[43] D. Fuchs and S. Tabachnikov. *Mathematical Omnibus: thirty lectures on classical mathematics.* American Mathematical Society, 2007.

[44] G. Cardano, *Ars Magna, sive de Regulis Algebraicis*, 1545, Courtesy of The Linda Hall Library of Science, Engineering & Technology.

[45] L. Gamwell. *Mathematics + art; a cultural history.* Princeton University Press, 2016.

[46] M.J. Gander and G. Wanner. From Euler, Ritz and Galerkin to modern computing. *SIAM Review*, 54(4), 2012.

[47] C.F. Gauss. *Disquisitiones generales circa superficies curvas.* Göttingen, 1827.

[48] M. Gessen. *Perfect Rigor: A Genius and the Mathematical Breakthrough of the Century.* Houghton Mifflin, 2009.

[49] P.J. Giblin. *Graphs, surfaces and homology: an introduction to algebraic topology.* Chapman and Hall, 1977.

[50] J.V. Grabiner. The changing concept of change: The derivative from Fermat to Weierstrass. *Mathematics Magazine*, 56(4):195–206, 1983.

[51] M.J. Greenberg. *Euclidean and non-Euclidean geometries; development and history.* W.H. Freeman and Co., 1993.

[52] R.D. Gregory. *Classical Mechanics.* Cambridge University Press, 2011.

[53] E. Hairer and G. Wanner. *Analysis by Its History.* Springer Undergraduate Texts in Mathematics, 2008.

[54] R.S. Hamilton. Three-manifolds with positive Ricci curvature. *J. Differential Geometry*, (17):255–306, 1982.

[55] F. Hartmann and R. Jantzen. Apollonius's ellipse and evolute revisited - the alternative approach to the evolute. *Convergence*, (Aug.), 2010.

[56] H. Hattori. *Partial differential equations; methods, applications and theories.* World Scientific, 2013.

[57] J. Havil. *Curves for the mathematically curious*. Princeton University Press, 2019.

[58] K.C. Howell. Three-dimensional periodic halo orbits. *Celestial Mechanics*, 32(1):53–71, 1984.

[59] D.H. Hubel and T.N. Wiesel. Receptive fields, binocular interaction and functional architecture of a cat's visual cortex. *Journal of Physiology*, (160):106–154, 1962.

[60] M. Hughes. How to be more Pythagoras. *Chalkdust magazine*, (17):46–49, 2023.

[61] Image in the public domain.

[62] Isaac Newton; Translated into English by Andrew Motte in 1729. *Sir Isaac Newton's Mathematical Principles of Natural Philosophy and his System of the World; translation revised by Florian Cajori*. University of California Press, 1966.

[63] J.-I. Itoh and K. Kiyohara. The cut loci and the conjugate loci on ellipsoids. *Manuscripta Mathematica*, 114(2):247–264, 2004.

[64] F. John. *Partial Differential Equations*. Springer-Verlag, 1982.

[65] C.R. Jordan and D.A. Jordan. *Groups*. Edward Arnold, 1994.

[66] D.W. Jordan and P. Smith. *Nonlinear ordinary differential equations; an introduction to dynamical systems*. Oxford University Press, 1999.

[67] J.V. José and E.J. Saletan. *Classical dynamics; a contemporary approach*. Cambridge University Press, 1998.

[68] V.J. Katz. *A history of mathematics; an introduction*. Addison-Wesley, 1998.

[69] K. Kitagawa and T. Revell. *The Secret Lives of Numbers; A Global History of Mathematics & its Unsung Trailblazers*. Viking, 2023.

[70] W. Klingenberg. *Riemannian Geometry*. De Gruyter Studies in Mathematics, 1995.

[71] M. Kuga. *Galois' dream: group theory and differential equations*. Birkhäuser, 1993.

[72] S. Lang. *Algebra*. Springer, 2002.

[73] P.-S. Laplace. *Traité de mécanique céleste*. Duprat, Paris, 1799.

[74] G. Leibniz. Nova methodus pro maximis et minimis, itemque tangentibus, quae nec fractas nec irrationales quantitates moratur, et singulare pro illis calculi genus. *Acta Eruditorum*, pages 467–473, 1684.

[75] T.-Y. Li and J.A. Yorke. Period three implies chaos. *The American Mathematical Monthly*, 82(10):985–992, 1975.

[76] S. Lloyd. Least squares quantization in PCM. *IEEE Transactions of Information Theory*, 28(2):129–137, 1982.

[77] E.S. Loomis. *The Pythagorean proposition*. Tarquin Publications, 1968.

[78] B.B. Mandelbrot. *Fractals: form, chance, and dimension*. W.H.Freeman and Co., 1977.

[79] B.B. Mandelbrot. *The Fractal Geometry of Nature*. W.H.Freeman and Co., 1982.

[80] A. McRobie. *The seduction of curves; the lines of beauty that connect mathematics, art, and the nude*. Princeton University Press, 2017.

[81] J. Milnor. *Topology from the differentiable viewpoint*. The University Press of Virginia, 1981.

[82] J. Milnor. Towards the Poincaré conjecture and the classification of 3-manifolds. *Notices of the AMS*, 50(10):1226–1233, 2003.

[83] M. Mirzakhani. Growth of the number of simple closed geodesics on hyperbolic surfaces. *Annals of Mathematics*, (168):97–125, 2008.

[84] C.W. Misner, K.S. Thorne, and J.A. Wheeler. *Gravitation*. W.H. Freeman and Company, 1970.

[85] C.D. Murray and S.F. Dermott. *Solar System Dynamics*. Cambridge University Press, 1999.

[86] P.J. Nahin. *When least is best*. Princeton University Press, 2004.

[87] A. Okrent. *In the land of invented languages.* Random House, 2010.

[88] Only joking, but wouldn't it be amazing?

[89] S. Ornes. *Math Art: Truth, Beauty and Equations.* Sterling New York, 2019.

[90] E. Outerelo and J.M. Ruiz. *Mapping Degree Theory.* AMS and Real Sociedad Matemàtica Española, 2009.

[91] A. Pannekoek. The planetary theory of Laplace. *Popular Astronomy,* 56:300.

[92] G. Perelman. Ricci flow with surgery on three-manifolds. https://arxiv.org/abs/math/0303109.

[93] Photo taken by the author.

[94] H. Poincaré. Mémoire sur les courbes définies par une équation différentielle. *Journal de mathématiques pures et appliquées,* 7:375–422, 1881.

[95] H. Poincaré. Analysis situs. *J. de l'École Polytechnique,* 2(1):1–123, 1895.

[96] D.S. Richeson. *Euler's gem: the polyhedron formula and the birth of topology.* Princeton University Press, 2012.

[97] B. Riemann. *Ueber die Hypothesen, welche der Geometrie zu Grunde liegen. In: Abhandlungen der Königlichen Gesellschaft der Wissenschaften zu Göttingen 13, S.133-150.* 1868.

[98] Rowe and McCleary. *The history of modern mathematics, volume II: institutions and applications.* Academic Press, 1989.

[99] S. Sahu and J. Foster. A continuum model for lithium plating and dendrite formation in lithium-ion batteries: formulation and validation against experiment. *Journal of Energy Storage,* (60):106516, 2023.

[100] I.M. Serrano and B.D. Suceavă. A medieval mystery: Nicole Oresme's concept of *curvitas. Notices of the AMS,* 62(9):1030–1034, 2015.

[101] D. Sheard. Apollonian packing. *Chalkdust magazine*, (11):36–43, 2020.

[102] D. Siersma. Poincaré and analysis situs, the beginning of algebraic topology. *Nieuw archief voor wiskunde*, 13(5):196–200, 2012.

[103] G. F. Simmons. *Differential Equations; with applications and historical notes.* McGraw-Hill Book Company, 1972.

[104] S. Singh. *Fermat's Last Theorem.* Fourth Estate, 1997.

[105] M. Somerville. *Mechanism of the heavens.* J. Murray, 1831.

[106] M. Somerville and M.C. Somerville. *Personal recollections, from early life to old age, of Mary Somerville : with selections from her correspondence.* J. Murray, 1873.

[107] D. Speiser. The Kepler problem from Newton to Bernoulli. *Archive for History of Exact Sciences*, 50(2):103–116, 1996.

[108] B. Stenhouse. Finding mathematical community; a reflection on the work of Mary Somerville. *Mathematics Today*, 58(6):192–195, 2022.

[109] J. Stillwell. *Mathematics and its history.* Springer, 2010.

[110] J. Stillwell. *Papers on Topology: Analysis Situs and Its Five Supplements.* AMS and London Mathematical Society, 2010.

[111] G. Strang. *Introduction to Linear Algebra.* Wellesley - Cambridge Press, 2016.

[112] S. Strogatz. *Nonlinear Dynamics And Chaos: With Applications To Physics, Biology, Chemistry, And Engineering.* CRC Press, 2000.

[113] S. Strogatz. *Infinite powers; the story of calculus, the language of the universe.* Atlantic Books London, 2019.

[114] D.J. Struik. *A Concise History of Mathematics.* Dover Publications, 1987.

[115] E.F. Taylor and J.A. Wheeler. *Spacetime Physics; introduction to special relativity.* W.H. Freeman and Co., 1992.

[116] W.P. Thurston. *Three-dimensional geometry and topology; volume 1.* Princeton University Press, 1997.

[117] E. Tupes and R. Christal. Stability of personality trait rating factors obtained under diverse conditions. *Personnel Laboratory, Wright Air Development Centre, Air Research and Development Command, United States Air Force,* 1958.

[118] B. van Brunt. *The calculus of variations.* Springer, 2004.

[119] B.L. van der Waerden. *A history of algebra; from al-Khwārizmī to Emmy Noether.* Springer-Verlag, 1985.

[120] M. Vermeeren. The power of curly brackets. *Chalkdust magazine,* (15):29–35, 2022.

[121] P. Verrier, T.J. Waters, and J. Sieber. Evolution of the $\mathcal{L}_1$ halo family in the radial solar sail circular restricted three-body problem. *Celestial Mechanics and Dynamical Astronomy,* 120:373–400, 2014.

[122] Viète F., *Opera Mathematica,* 1646, Courtesy of The Linda Hall Library of Science, Engineering & Technology.

[123] M.D. Waller. Vibrations of free square plates: part I. Normal vibrating modes. *Proc. Phys. Soc.,* (51):831–844, 1939.

[124] T.J. Waters. Regular and irregular geodesics on spherical harmonic surfaces. *Physica D: Nonlinear Phenomena,* 241:543–552, 2012.

[125] T.J. Waters. Integrable geodesic flows on tubular sub-manifolds. *Proceedings of the International Geometry Center,* 10(3-4):17–28, 2018.

[126] T.J. Waters. The conjugate locus on convex surfaces. *Geometriae Dedicata,* 200:241–254, 2019.

[127] T.J. Waters and M. Cherrie. The normal map as a vector field. *Balkan journal of geometry and its applications,* 27(2):145–157, 2022.

[128] T.J. Waters and M. Cherrie. The conjugate locus in convex 3-manifolds. *New Zealand Journal of Mathematics,* 54:17–30, 2023.

[129] T.J. Waters and B. Nolan. Gauge invariant perturbations of self-similar Lemaître-Tolman-Bondi spacetime: even parity modes with $l \geq 2$. *Physical Review D*, 79(8):084002 1, 2009.

[130] J.R. Weeks. *The Shape of Space*. CRC Press, 2002.

[131] T. Whitmarsh. *Battling the gods; atheism in the ancient world*. Faber & Faber, 2017.

[132] Why does this balloon have -1 holes? (Stand-up Maths). `https://www.youtube.com/watch?v=ymF1bp-qrjU`

[133] With permission from Diane Davies, Maya gods and religious beliefs. `https://www.mayaarchaeologist.co.uk/public-resources/maya-world/maya-gods-religious-beliefs/`

[134] With permission from J.R. Weeks, Curved Spaces `https://www.geometrygames.org/`

[135] With permission from The Coding Train `https://thecodingtrain.com/`

[136] With permission from The Royal Belgian Institute of Natural Sciences, Brussels.

[137] J.G. Yoder. *Unrolling Time: Christiaan Huygens and the Mathematization of Nature*. Cambridge University Press, 1989.

Index

Printed in the United States
by Baker & Taylor Publisher Services